OFF GRID AND MOBILE SOLAR POWER FOR EVERYONE

Your Smart Solar Guide

by Lacho Pop, MSE and
Dimi Avram, MSE

Digital Publishing

Disclaimer Notice

The authors of this ebook, named '**Off Grid And Mobile Solar Power For Everyone: Your Smart Solar Guide**', hereinafter referred to as the 'Book', make no representation or warranties with respect to the accuracy, applicability, fitness, or completeness of the contents of the Book. The information contained in the Book is strictly for educational purposes.

Summaries, strategies, tips and tricks are only recommendations by the authors and the reading of the Book does not guarantee that reader's results shall exactly match the authors' results.

The authors of the Book have made all reasonable efforts to provide current and accurate information for the readers of the Book and the authors shall not be held liable for any unintentional errors or omissions that may be found.

The Book is not intended to replace or substitute any advice from a qualified technician, solar installer or any other professional and advisor, nor should it be construed, as legal or professional advice and the authors explicitly disclaim any responsibilities for such use.

The installation of solar power systems requires certain background professional qualification and certification for working with high voltages and currents dangerous to human life and for installing solar power systems and appliances. The reader should consult every step of their project or installation with a qualified solar professional, installer or technician and local authorities.

The authors shall in no event be held liable to any party for any direct, indirect, punitive, special, incidental or other consequential damages arising directly or indirectly from any use of this Book, which is provided on "as is, where is" basis, and without warranties.

The use of product names, brand names and trademarks in the Book doesn't imply any endorsement.

About the Authors

Lacho Pop, MSE, has more than 20 years of experience in market research, technological research and design and implementation of various sophisticated electronic and telecommunication systems. His large experience helps him present the complex world of solar energy in a manner that is both practical and easy understood by a broad audience. He authored and co-authored several practical solar books in the field of solar power and solar photovoltaics. All the books were well received by the public.

Dimi Avram, MSE, has more than 15 years of experience in engineering of electrical and electronic equipment. He has specialized in testing electronic equipment and performing techno-economic evaluation of various kinds of electric systems. His excellent presentation skills help him explain even the most complex stuff to anybody interested. He co-authored several practical solar books in the field of solar power and solar photovoltaics. All the books were well received by the public.

You may contact the authors by visiting the website: solarpanelsvenue.com ,
or by email: author@solarpanelsvenue.com

Also by the authors:

The Ultimate Solar Power Design Guide: Less Theory More Practice (The Missing Guide For Proven Simple Fast Sizing Of Solar Electricity Systems For Your Home, Vehicle, Boat or Business)

[Paperback and all type of eReaders editions – Kindle, Kobo, Nook, Apple, etc.]

ISBN-10: 6197258048
ISBN-13: 978-6197258042

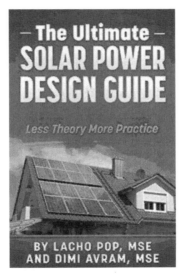

The Ultimate Solar Power Design Guide is a straightforward step by step guide on solar power system sizing. It is written by experts for beginners and professionals alike.

The New Simple And Practical Solar Component Guide [Paperback and all types of eReaders editions – Kindle, Kobo, Nook, Apple, etc.]

ISBN-10: 6197258099
ISBN-13: 978-6197258097

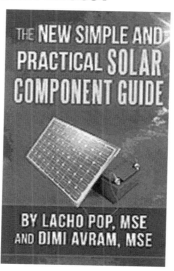

The book "**The New Simple And Practical Solar Component Guide" is a simple and practical guide** that demystifies all of the components of a solar power system in a way that anyone lacking technical background can understand. By reading this book you will get a step-by-step know-how to select and combine solar components of your solar power system to get the most efficiency for lower price.

The Truth About Solar Panels: The Book That Solar Manufacturers, Vendors, Installers And DIY Scammers Don't Want You To Read [Paperback and all types of eReaders editions – Kindle, Kobo, Nook, Apple, etc.]

ISBN-10: 6197258013
ISBN-13: 978-6197258011

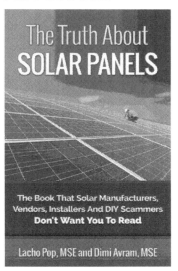

The book is about solar photovoltaic panels – the main building unit of solar systems. By reading it you get a complete know-how to buy, build, compare, evaluate, mix and assemble different type of solar panels.

Top 40 Costly Mistakes Solar Newbies Make: Your Smart Guide to Solar Powered Home and Business 2016 Edition [Kindle and Paperback Edition], Kindle ASIN: B01GGB7QP8, Paperback ISBN-13: 978-6197258073

The book "Top 40 Costly Mistakes Solar Newbies Make" is a simple and practical guide that could save you a lot of money, headaches and time during the planning, buying, implementation, and operation phase of your solar power system.

Table of contents

xiv

Introduction

This book is written for solar power enthusiasts making their first steps in the world of solar photovoltaic energy. Here are the essentials of off-grid solar panel systems revealed.

The book, however, is also targeted to intermediate and advanced solar users due to its systematic and simplified step-by-step approach to solar system design. The essentials of off-grid and mobile solar panel systems are explained in an easy-to-follow-and-grasp manner. Furthermore, every solar enthusiast, regardless of their experience level, will benefit from the essential information revealing how to scale up fast, easy and cost-effectively a solar power system.

Off-grid solar electric systems are not connected to the electricity grid. They are preferred in remote areas where buildings are far from any utility infrastructure.

Although mobile solar systems are a subset of off-grid solar, they do require some particular attention and do have their specifics when it comes to component selection and system design. All these details are carefully noted and explained in the book. What is more, a whole chapter devoted to mobile power system sizing explains in a step-by-step manner how to design a system that is both best-performing and cost-effective.

As a rule, in off-grid solar systems, the produced electricity is matched to the user's energy needs. A typical off-grid system requires an electricity storage system because electrical energy might not be needed at the same time when generated. For this reason, battery banks are used for electricity storage, and most off-grid photovoltaic systems are battery-based. The other option for power backup is a hybrid solar system where the capacity of the electricity storage can be enhanced by adding an extra power source – for example, a diesel generator.

In areas located away from any existing utility poles, it is often less expensive to install an off-grid photovoltaic power system than pay a fortune to get connected to the grid. Therefore, off-grid solar panel systems are a good solution for households located away from the public electricity grid and business clients in industrialized countries as an option for lower-cost production methods.

It should be noted that not all the household might be dependant on solar-generated electricity. A solar power system might be an option for individual devices and applications – lamps, pumps, SOS telephones, traffic signal systems, ticket machines, clocks, solar radios, mobile and measuring equipment on cars and vans, etc.

Off-grid solar panel systems are attractive, as they:

- Are a reliable source of power
- Are practically unlimited in size – an off-grid solar power system could serve a single device or a couple of buildings with a complex electrical network
- Can be installed almost anywhere in the world
- Can be less expensive than paying for getting connected to the utility grid
- Can provide electricity to the most household devices
- Can be used on vehicles – RV and marine.

Here are some of the primary applications of off-grid solar panel systems:

- Small solar systems for household use
- Providing power to remote homes or summer villas
- Recreational vehicles
- Water pumps and cooling
- Power supply for telecommunication equipment
- Street infrastructure equipment – street lamps, bus station dashboards, parking meters, etc.
- Remote meteorological stations and airports.

The purpose of the authors is to provide you with all the practical information you need to build an off-grid solar system fast and easy by yourself.

Furthermore, you are going to acquire the essential knowledge to make you confident when reading technical documentation, communicating with solar vendors and installers, and selecting the components of the off-grid system configuration that fits best your needs and available budget.

The sizing formulas are simplified in a way which does not sacrifice their accuracy to be of help for those who just cannot get started or hate mathematics. A more sophisticated version of these formulas is given in the appendices at the end of the book.

How should you use this book to get the most out of it?

We recommend you first to read the chapters related to the basics of electricity and solar power. If you feel that your knowledge of this matter is enough, you could just skim through it to fill any possible gaps in your understanding. The schematic drawings will help you not only grasp the information but also immediately spot such gaps. If you think you have enough knowledge about the subject, please, go to the next chapter.

After reviewing the electricity and solar basics, and being aware of the essentials of the system components, you can go on with the chapters revealing you how to size an off-grid and a mobile solar power system in a step-by-step manner. The design chapters contain simplified yet accurate mathematical equations on sizing and assembling a solar power system for your home or vehicle. Eventually, you will get a system that is reliable, cost-efficient, optimized for maximum performance and easy to maintain.

If you think that your solar case requires more accurate calculations, please, refer to the corresponding appendices at the end of the book. There you can find some more complicated yet more accurate formulas on system sizing.

We recommend you to check both sizing examples concerning the design of an off-grid or a mobile solar power system, regardless of the particular configuration you are planning to design. All the necessary system design information is gradually introduced, with the two sizing examples complementing each other.

We wish you lot of success in your solar endeavor!

Electricity basics

Current

Electrical current is the rate of the flow of free electrons moving directionally within a conductor. Electrons can move directionally when affected by voltage.

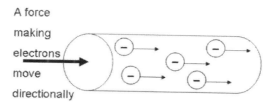

Electrical current can only flow through materials containing free electrons. In general, the materials are divided into:

> **Conductors** – all the metals. They contain free electrons, so they can pass the electric current through.

> **Insulators** – rubber, paper, plastic, wood, etc. They do not contain free electrons, so they stop the electric current.

> **Semiconductors** – silicon, gallium arsenide, etc. Free electrons are only available upon specific favorable conditions – temperature, sunlight, human touch, etc.

Important!

Photovoltaic solar panels are made of semiconductor, usually silicon.

In a solar panel, there are free electrons and current can flow through upon sunlight applied to the surface of the panel:

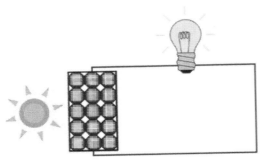

Electrical current can only flow upon voltage applied. Electrical current is denoted by the letter 'I' and is measured in **amperes** or **amps** (**A**) by a device called '**ammeter**.'

Voltage

Voltage is the force of pressure making the free electrons move directionally. Voltage can only appear between two points of different potential energy. For a battery cell (further explained), such two points are the positive and the negative poles:

'Potential' energy means 'hidden' energy capable of performing some work. The difference between the 'hidden' or 'potential' energy of two different locations gives drive to the electrons, and unless they meet some obstacles on the way, they start moving directionally. Thus, the available potential energy is transformed into motion.

It is the voltage that makes the electrical current flow. The direction of the current is adopted to be the opposite to the direction of the moving electrons.

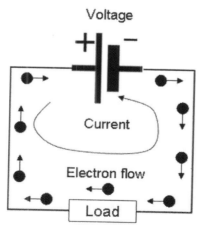

21

While in motion, the moving electrons render their energy to both the connecting cables and the plugged-in loads (electrical devices and appliances). When the electrons release their energy, the cables get heated and the loads perform some work – you have the lights on, watch TV or play computer games.

Voltage is denoted by the letter '**V**' and is measured in **Volts** (V) by a device called '**voltmeter**.' Often current and voltage are measured by a universal measuring device called 'multimeter.'

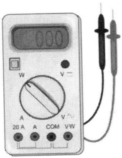

Multimeters measure both direct and alternating electricity (refer to the next section).

'Direct' and 'alternating' electricity refers both to current and voltage. 'Direct' means that the value and the direction of the current remain constant over time.

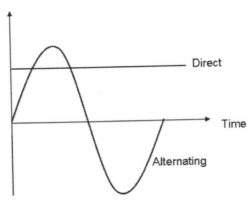

'Alternating' means that the value and the direction of the current change in time according to a law.

The most common law is the sinusoidal law where the AC current changes its value over time according to a sinusoid (pictured above). The result is a sinusoidal ('sine') wave.

The positive and negative poles of DC electricity sources (also known as 'batteries') are always pictured, since their position is fixed:

On the previous picture, you see that the sinusoid goes above and below zero. Therefore, the positive and negative poles of an AC generator are not fixed – their positions change in time.

Important!

An essential difference between DC and AC electricity:

DC electricity can be stored in a battery for later use, unlike AC electricity that cannot be stored and should be consumed right away.

Alternating electricity is available in households connected to the local electrical grid.

AC electricity:

- Provided by public utility companies
- Produced by fuel generators
- Cannot be stored for later use.

DC electricity:

- The current and the voltage have a constant direction.
- Produced by batteries.
- Can be stored for later use.
- Widely used by lots of devices (radios, radio clocks, flashlights, water pumps, DC fans, DC fridges, electrical shavers, etc.)

Important!

Solar panels produce DC electricity. Such solar-generated electricity can either be consumed immediately or stored in a battery for later use.

What can be done with the DC electricity generated by solar panels, provided that most of our household devices operate on AC?

DC is converted into AC by a device called '**inverter**.' Since solar panels only generate DC electricity, the inverter is a must if you have devices operating on AC.

Let's compare voltage and current:

- Current can start flowing only if there is voltage applied,
- Both voltage and current have direction,
- DC voltage makes DC current flow, while AC voltage makes AC current flow.

Important!

When buying a device, you should be careful about:

- The voltage type (AC or DC),

- The rated voltage (120V or 230V for AC; 12V, 24V or 48V for DC).

Otherwise, that device either will get damaged or will not work!

Power

Upon voltage applied between two points, current starts flowing through, and electrical power is produced.

Electrical power is the multiple of the voltage and the current. Power is denoted by the letter '**P**' and is measured in Watts:

Power [in watts] = voltage [in volts] * [current in amps]
P [W] = V [V] * I [A]

The above formula shows the relation between the power, the voltage, and the current. Here is the message, 'When either voltage or current increases, power goes up too.'

Example: *If you measure a voltage of 24 Volts and a current of 0.5 amps on an electrical load, the power rendered on such a load is 24 * 0.5 = 12 Watts.*

Important!

Power can be either consumed (used) or generated (produced). Generators produce power, while electrical loads (the devices in your household or office) consume power.

Photovoltaics are generators of electrical power.

A solar panel produces DC electricity, while a battery stores DC electricity. Nevertheless, both of them are sources of power, since they produce voltage which makes current flow.

Power is stated on the label of every electrical device. This stated power is often called 'rated power' or 'power rating.'

Energy

Energy is the electricity either used or produced over a period. Physically, it is the useful work done by electrons moving directionally along a conductor during 1 hour.

To differentiate between power and energy, energy is the work performed, while power is the rate of performing work.

Energy is measured in Watt-hours [Wh]. 1 Wh of energy is the power produced for 1 hour of work. Another measurement unit for energy is kilowatt-hours (kWh). 1 kWh of energy, or 1,000 watts per hour, is the power produced during 1,000 hours of work.

Here is the basic energy law:

E [Watt-hours] = P [Watts] * Time [hours].

Since

P [Watts] = V [Volts] * I [Amps],

The energy can also be expressed as:

E [Watt-hours] = V [Volts] * I [Amps] * Time [hours]

Example: Your TV set is rated 200 W and you want to watch your favorite team in a match lasting 2 hours. Therefore, your TV has consumed 200 Watt * 2 hours = 400 Watt-hours (Wh) of electrical energy. To power your TV, your solar system should produce 400 Wh of solar-generated electricity.

Important!

Power is the instant electricity needed for the devices used at a given moment. The more devices plugged in, the more power consumed. Energy refers not to an instant of time but rather to a longer period. Although measured in watt-hours, energy measures the electricity you need if you use these devices for some longer – a day, a month or a year.

'Solar-generated electricity' means 'the energy generated by the solar system'.

Electric circuits

An electric circuit is a fixed, closed path where electric current can flow through a conductor (cables) and loads, upon voltage (generated by battery, solar panel or AC generator) applied.

If you have a solar panel as a power source and connect it to a load by cables, you have a closed circuit. In this closed circuit, upon exposure to direct sunlight, the solar panel generates voltage **V** causing the current **I** to flow:

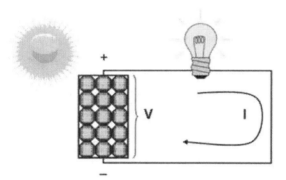

Short circuit

A 'short circuit' occurs when the two terminals (poles) of a voltage source are directly connected.

As you can see, in a short circuit the load is missing. The moving electrons have nowhere to render their energy. Since they do not meet any resistance along their way, the current boosts dramatically. As cables are not electrical devices, electrons cannot do much useful work with them. All the electrons can do is release their energy as heat. If all the energy is converted into heat, the cables get overheated and fire is likely to occur.

In other words, when short-circuited, the electrical generator can get damaged. For a lead-acid battery bank (revealed in the 'Batteries' chapter), this might even result in an explosion. In case of short circuit, the current reaches its maximum.

Open circuit and voltage drops

If you connect a voltmeter to the poles of a battery, the voltmeter will show the battery's nominal (rated) voltage.

No current can flow through a voltmeter. This is called an 'open circuit':

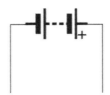

In an open circuit:

- **V=max**
- **I=0.**

When you plug the battery into a circuit, and add loads and cables, you close the circuit and the current starts flowing. Each part of a closed circuit creates a **voltage drop**. Here is what is valid for any closed circuit:

- A voltage drop has a value always lower than the battery voltage.
- The source voltage is the sum of all voltage drops created.

Loads are no exception; they create voltage drops too. Unfortunately, voltage drops are also created on the cabling, which is undesired. Voltage drops on the cabling are known as 'cable losses.'

The higher the voltage drop on the cabling, the lower the voltage drop on the load. Unfortunately, a part of the voltage is always inevitably lost on the cables.

Important!

In case of a too high voltage drop on the cables, the voltage drop on the load can turn out to be insufficient. Therefore, the load might not be able to operate at all.

In such a situation, there are two options available:

- Choose a battery of a higher voltage. The cable losses will remain high, but higher voltage will be delivered to the load as well.
- Keep the battery, but replace the long cables with shorter ones, of lower losses, so that

enough voltage is delivered to the load.

The sum of the voltage drops in a circuit is constant. It is equal to the rated voltage of the power source.

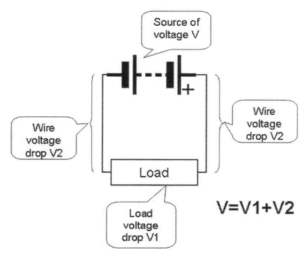

Let's see an example.

If you use longer cables, they will produce a higher voltage drop **V2**. Since the voltage **V** of the source is constant, if **V2** increases (according to the formula **V=V1+V2**), the voltage drop **V1** on the load decreases too.

Therefore, the load might not be able to operate, since it cannot get the voltage it needs.

What does this mean? When you add a new voltage drop in a closed circuit, all the existing voltage drops decrease, since the sum of all the voltage drops available is supposed to be the same.

Resistance

The cause of voltage drops is called 'resistance'. Resistance is the ability of a material to resist to current. In other words, 'resistance' denotes how difficult it is for the current to flow across a circuit. The higher the resistance, the lower the current that can flow.

Conductors have low resistance – electrons do not experience any serious problems during their directional movement. On the opposite, insulators have a very high resistance – no electrons can move within, and no current can flow through. Semiconductors are in the middle – under certain conditions, their resistance is high, and no current can flow through. Under other conditions – for example, when exposed to sunlight – the resistance decreases dramatically, and semiconductors become conductors – the electrons 'get free.'

Heat is another tangible effect of resistance. When electrons flow through a material and meet certain resistance, they always render some part of their energy, and the material gets heated. Typically, heat is dissipated around in the air. When a piece of cable gets more heat that it can dissipate, fire is likely to occur, as is the case of short circuit.

Resistance can be expressed as a relation between voltage and current, called the 'Ohm's law' – the basic law of the electrical engineering:

R= V ÷ I,

Or expressed otherwise:

V = R * I

Resistance is measured in Ohms by a device called 'ohmmeter.' An ohmmeter is often a part of a multimeter.

Important!

Resistance depends on the cable's:

- Length – the longer the piece of material, the higher the resistance. Longer cables have higher resistance and hence, create higher losses.

- Cross-section area (diameter) – the smaller the cross-section, the higher the resistance, since the current is lower. If you want to reduce the cable losses, you should select cables of a larger cross-section, which will cost you money.

Systems for electricity generation are called 'power systems.' Therefore, any solar panel photovoltaic system is a power system. In power systems, electricity is generated by generators and delivered to loads. If you have a solar panel system, it delivers the electricity generated by the solar panels on the roof to the electrical devices in your room.

Due to the cable losses, transferring the electricity generated is never 100% efficient. Transmission losses are higher in DC power systems than in AC power systems. The battle is always for lower cable losses. In most of the cases, this means using shorter cables.

Important!

If you use longer cables, you need to:

- Either use a DC source of higher voltage (which is not always possible), or

- Use cables of a larger cross section, which are more expensive.

Obviously, both of the above – buying a higher voltage

solar panel or battery or using cables of a larger cross-section – mean increasing the system implementation costs.

Connecting loads

You can connect many loads to a circuit. There are two main types of connecting loads – in series and parallel.

Loads connected in series:

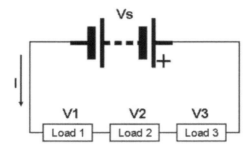

When loads are connected in series, there is the same current flowing through each load. The total voltage is a sum of the voltage drops on each load:

Vtotal = V1 + V2 + V3

Loads connected in parallel:

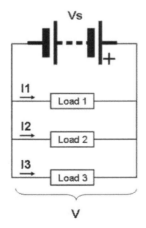

Upon loads connected in parallel, there is the same voltage applied to each load. The total current in the circuit is a sum of the currents flowing through each load:

Itotal = I1 + I2 + I3

It should be noted that both in series and parallel connection, the total power is the same:

Ptotal_series = Vtotal * I = P1 + P2 + P3
Ptotal_parallel = V * Itotal = P1 + P2 + P3

Important!

Which connection is preferable – in series or parallel?

Connecting loads in series is not recommended because:

1) If a load fails, current cannot flow through, and other loads cannot operate either.

2) Adding a new load is problematic. Since adding a new load means creating a new voltage drop, in case of too many voltage drops, the battery voltage might not be sufficient to power all the loads.

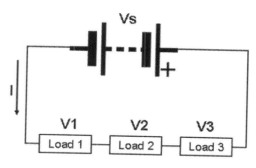

Important!

Connecting loads in parallel is recommended, as:

- Each load is independent of the others. If a load fails, the other loads continue to operate, as the current only stops flowing through the failed load.

- Adding new loads is not a problem – the other loads are not affected, since in parallel connection, voltage is the same on all the loads.

For this reasons, ALL the electrical devices at your home or office are connected in parallel.

Loads in solar power systems

Loads directly affect the performance of solar power systems.

Extra loads can cause a system to fail if they require more power than the solar panels can generate or the battery bank can store. Furthermore, the efficiency of the loads impacts the solar system's performance. Therefore, the loads connected to a solar panel system should be as efficient as possible. Oversized or improper loads often result in a system failure. A basic rule for any solar power system is that you should not use a device unless you really need it.

Example:

Here is an example list of daily loads:

Device	Rated power	Period of use	Energy consumed
TV set	100 W	3 h	300 Wh
Laptop	60 W	8 h	480 Wh
Lights	15 W	12 h	180 Wh
Radio	40 W	10 h	400 Wh
Vacuum cleaner	700 W	0.5 h	350 Wh
Total energy consumed:			**1,710 Wh**

If this is the regular everyday consumption, then every month the user has to pay for 1,710 Wh * 30 days = 51,300 Wh or 51.3 kWh of energy consumed.

On the next page, you can find the typical power consumption of some commonly used devices.

Typical power consumption of some household appliances

Source: http://www.wholesalesolar.com

Appliance	Wattage
Laptop	60-250
LCD TV	213
Plasma TV	339
12" black and white TV	20
19" color TV	70
25" color TV	150
Satellite dish	30
Cell Phone - recharge	2-4 watts
Computer	120
Monitor	150
Standard TV	188
Stereo	60
Video Game Player	195
MP3 Player - recharge	0.25-0.40 watts
Radiotelephone - Receive	5
Radiotelephone - Transmit	40-150
Clock Radio	7
Blender	300
Cable Box	20
Can Opener	100
Ceiling Fan	100
Central Air Conditioner	5
Coffee Machine	1,5
Curling Iron	90
Dehumidifier	350
Humidifier	300-1000
Dishwasher	1200-1500
Electric blanket	200
Espresso Machine	360
Hair Dryer	1,538
Shaver	15
Vacuum Cleaner	500
Microwave	1,5
Refrigerator/ Freezer	
16 cu. ft. (AC)	1200 Wh/day
20 cu. ft. (AC)	1411 Wh/day
Freezer	
15 cu. ft. (Chest)	1080 Wh/day
15 cu. ft. (Upright)	1240 Wh/day

Appliance	Wattage
Hedge trimmer	450
1/2" drill	750
1/4" drill	250
1" drill	1000
12" chain saw	1100
14" band saw	1100
Waterpik	100
Weed eater	500
Well Pump (1/3-1 HP)	480-1200
3" belt sander	1000
7-1/4" circular saw	900
8-1/4" circular saw	1400
Heaters	
Electric Clothes Dryer	3,4
Engine Block Heater	150-1000
Stock Tank Heater	100
Furnace Blower	300-1000
Hot Plate	1200
Iron	1,1
Portable Heater	1500
Toaster	1,1
Toaster oven	1,2
Water heater	479
Room Air Conditioner	1,1
Lights	
100 watt incandescent bulb	100
20 watt DC compact fluor.	22
25 watt compact fluor. bulb	28
CFL Bulb (60-watt equivalent)	18
CFL Bulb (75-watt equivalent)	20
40 watt DC halogen	40
50 watt DC incandescent	50

DC loads

Lighting (incandescent or quartz halogen bulbs) and other resistive loads can be powered by solar-generated electricity. A lamp and its fixture should be as efficient as possible. Using DC lighting equipment is a smart way to bypass the inefficiency related to the DC-AC conversion.

"Heating" loads are not recommended to be powered by solar-generated electricity. "Heating" loads comprise resistive heating appliances and tools such as toasters, coffee makers, soldering irons, room heaters and water heaters. Due to the high energy consumed, such loads should only be used in case of no other option available.

Inductive loads contain a motor or electromagnet. Many solar electric systems are designed to provide power to DC motors driving tools, fans, pumps, and other appliances. The efficiency of an inductive load should be as high as possible. DC motors are noted for their higher efficiency compared to AC motors.

Electronic loads, such as various audio-visual and communication electronic equipment, devices for data collection or security, are typically DC-powered. These can be operated by solar electric systems with no problems. Many of these loads are sensitive to even small variations in voltage.

AC Loads

AC loads can only be used if the solar electric system includes an inverter. A right approach is to limit the number of the AC loads as much as possible due to the energy lost during the DC-AC conversion performed by the inverter.

With photovoltaic systems, AC incandescent lighting should be minimized because of its poor efficiency. For this reason, AC fluorescent lighting is recommended as more efficient. Furthermore, usage of AC appliances, such as toasters, dryers, soldering irons, and heaters should be minimized.

Inductive and electronic loads

Many appliances comprise motors operating on AC. Such devices typically require a "clean" source of AC power, i.e., an inverter of a pure sinusoidal waveform, which is more expensive. Motors powered by an "unclean" enough power source waste electrical energy dissipated through the motor housing as heat. Thus, the lifespan of the device is reduced. Other inductive types of load with a "clean" AC power supply are microwave ovens. A "clean" AC power supply generates a sine-wave voltage.

For some electronic devices, such as communication equipment and small computers, providing AC power by a simple inverter is usually enough. Other electronic devices, such as video and audio equipment, require advanced inverters.

Batteries are the solar system component that can be severely affected by the loads. If a solar electric system does not comprise a charge controller, any oversized loads or excessive use of loads can quickly damage the battery bank, and the latter will have to be replaced soon due to overdischarging. Another adverse effect is overcharging which happens during periods of few or no loads plugged-in at all, or during extended periods of abundant sunlight. For these reasons, the size of the battery bank must be tailored to the loads used. [2]

Energy efficiency

Achieving energy efficiency means reducing electrical consumption.

Important!

If you are planning to have an off-grid solar panel system, achieving energy efficiency is of the utmost importance!

Therefore, you should search to use as many energy-saving loads as possible.

Energy efficiency is vital because solar-generated electricity is still relatively expensive. By using energy efficient appliances, you will reduce the cost of your photovoltaic system.

The starting points for improving the energy efficiency include:

- Replacing the electrical devices consuming too much energy with non-electrical ones (working on gas, coal, etc.).

- Replacing any traditional devices with energy-saving ones (incandescent bulbs with fluorescent bulbs, etc.).

- Using DC devices (lamps, TV and radio sets, water pumps, drilling equipment).
- Turn off all standby loads (a.k.a. phantom loads), such as DVDs, TV sets and computers, while they are not in use.
- Upgrade your heating, ventilation, and air conditioning systems.
- Replace your old refrigerator and freezer with new, high-efficiency models (either energy-efficient or propane ones).
- Replace high-consumption loads (high-wattage electric stoves, clothes dryers, water heaters, heating appliances) with their propane or natural gas alternative versions. Certainly, a solar thermal water heating system is the perfect option.
- Replace electrical air-conditioning appliances with evaporating cooling systems.
- Make an external and/or an internal isolation of your building.
- Utilize your large electrical loads (pool pumps, electrical mills, etc.) during off-peak hours.
- Install solar water heaters and summer shadings.
- Utilize devices with timers and other home automation systems.

Off-grid system configurations

Off-grid solar panel systems are two types – standalone and hybrid.

Standalone systems are purely photovoltaic. Such systems only rely on solar energy to generate power. They are not backed by an additional source of electricity.

Hybrid systems are modified standalone systems provided with an alternative generator operating on wind, combustive fuel, etc.

Standalone systems are typically supplied with battery storage, since electricity might be needed when solar energy is not enough (or is lacking at all) to generate power, which can happen:

- At evening and night, or
- During periods of scarce sunlight (in winter or cloudy/rainy days).

When the daily electricity needs are too high, a solar panel system relying only on solar energy is not suitable. Otherwise, a too large and expensive battery bank is needed (which brings both high initial and maintenance costs), making the purely photovoltaic option far from cost-effective. In case of high daily power consumption, a hybrid off-grid system is preferred. Apart from the solar array, a hybrid system has an additional source of electricity, which supplements the battery storage.

Directly coupled standalone solar panel systems are the simplest and the most commonly used, as they:

- Only comprise a solar array, a battery (optional) and loads

- Can be used in a wide range of applications.

You can have the simplest standalone solar panel system by directly connecting a DC load to the solar array:

The load might be, for example, a DC fan or a DC water pump. Such devices use electricity right away after it is generated (i.e., in daytime), without any need to store it for later use.

Off-grid systems are typically provided with a battery backup to store the solar-generated electricity:

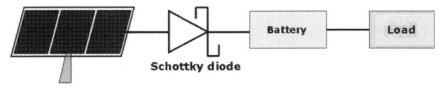

The load can be, for example, a TV set or a laptop. Since such devices operate not only in the daytime, a battery is needed to ensure their operation during night hours. The Schottky diode prevents the solar panels from the reverse current flowing from the battery to the solar panels during the night.

In such a system, the solar-generated electricity is used for charging the batteries, usually through a charge controller:

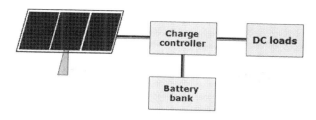

If the solar array has to power AC loads, an inverter is needed:

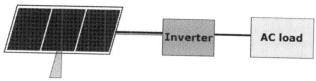

Inverters in standalone solar panel systems differ from inverters in grid-tied systems, although they apparently do the same – convert DC into AC electricity. A standalone inverter and a grid-tied inverter cannot be used interchangeably.

In standalone systems, the solar-generated electricity produced by the solar array must be enough to cover the total consumption of the loads plugged into the system.

Below you can find an example of a standalone system intended to replace the utility grid for remote locations:

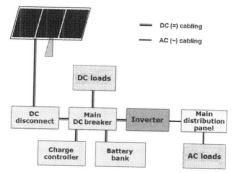

A simplified view of a stand-alone solar electric system

Here is a list of components of an off-grid standalone system:

- Photovoltaic array - generates DC electricity from sunlight.
- DC disconnect – disconnects the solar array from the rest of the system.
- Main DC breaker – connects the inverter to the battery and charge controller.
- DC loads – all devices operating on DC power.
- Charge controller – regulates battery charging, thus increasing the battery lifespan.
- Battery bank – stores the electricity generated by the solar array.
- Inverter – converts DC into AC electricity.
- Main distribution panel – protects the inverter from power surges and distributes the generated AC power to several branches of AC circuits. Usually, each branch has a standalone fuse or a circuit switch protection.
- AC loads – all devices consuming AC power.

Standalone systems are 'photovoltaic-only' systems. They contain no power generator other than the solar array.

The essentials of standalone systems

- Power efficiency is of utmost importance.
- Standalone systems cannot provide you with unlimited access to electricity.
- Standalone systems can take advantage of DC loads (lighting, refrigeration, electronic devices).

- Standalone systems are usually built in rural areas where you have more options for installing solar panels. Usually, solar systems in rural areas are installed on the ground rather than on a roof.

- Standalone systems are often built to perform optimally throughout the whole year. Since in winter the sun is located lower in the sky and nearer the horizon than in summer, the solar array needs a higher slope (tilt). Such a higher slope is much easier to achieve with a ground-mounted solar array.

- In a standalone system, the target is to provide all the facilities to store the solar-generated electricity for the sake of meeting the consumption needs at any moment. Often such a moment, however, does not coincide with the solar array's maximum power yield.

Limitations of standalone systems

- Unless your building is located too far from the utility grid, replacing the latter with a standalone solar panel system is not cost-effective. In such cases, a hybrid system might be the better option.

- Due to the solar radiation variability, a standalone solar system does not deliver a maximum performance all the year round. In winter, it is often more cost-effective to have a hybrid system than spend a fortune on a battery bank and rely solely on solar-generated electricity.

- The electricity produced by the solar array can be stored in batteries for limited periods only.

- Making your home energy efficient is a must before building a standalone system.

- In most residential standalone systems, you need a separate, well-ventilated room, as well as performing specific operation and maintenance activities for the battery bank.

The second type off-grid systems are **hybrid systems**.

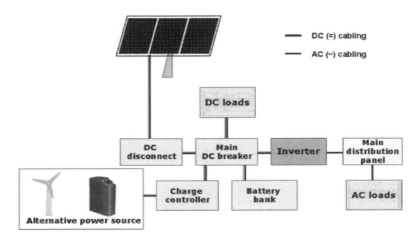

A simplified view of a hybrid off-grid system

A hybrid off-grid system is a standalone system with an alternative power source added – a wind generator or fuel generator.

Hybrid systems are preferred in cases of high energy consumption and/or extended periods of cloudy days, when a too bulky battery is needed. Such a battery is both costly and challenging to maintain.

What type of off-grid system to choose?

Whether to choose a standalone or a hybrid system depends on:

- Whether you use your building throughout the year or just on a seasonal basis.

- Whether your location is easily accessible.
- Your daily electricity consumption.
- What kind of devices you use – critical or not.

In a hybrid system, the alternative generator usually operates on diesel, propane or gasoline, and less commonly – by the wind.

Important!

As a rule, **hybrid systems are recommended when the daily electricity needs exceed 2.5 kW.**

Also, a hybrid system is always recommended when the available sunlight in your area is not enough to meet the desired days of autonomy, and this would result in a costlier battery bank.

Sources:

1. Antony, Falk, Christian Durschner, Karl-Heinz Remmers. 2007. Photovoltaics for Professionals: Solar Electric Systems Marketing, Design and Installation, Routledge.

2. Mayfield, Ryan. 2010. Photovoltaic Design and Installation for Dummies, Wiley Publishing Inc.

3. MSE Pop, Lacho, Dimi Avram MSE (2015-02-17), The New Simple and Practical Solar Component Guide (Kindle Locations 1198-1199). Digital Publishing Ltd.

Solar (photovoltaic) panels

An off-grid solar panel photovoltaic system converts the solar energy into electrical energy.

Photovoltaic (solar) modules are the main components of any photovoltaic system. A solar module consists of connected solar cells. Furthermore, to achieve a higher energy yield, solar modules are preassembled in panels, and panels, in turn, are connected in arrays.

In the picture below, you can distinguish between the main types of photovoltaic units.

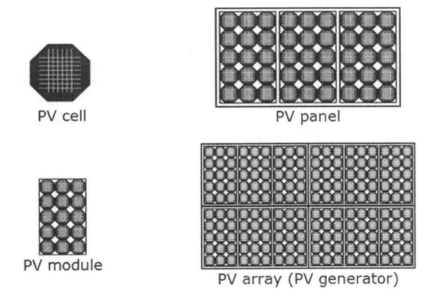

PV cell

PV panel

PV module

PV array (PV generator)

**PV Cell -> PV Module -> PV Panel
-> PV Array -> PV System**

Today 'solar module' and 'solar panel' are often used interchangeably.

Photovoltaic cells and modules are made of

semiconductor (usually silicon) capable of producing electricity when exposed to sunlight:

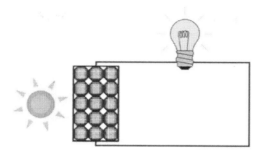

Such a capability is called 'photovoltaic effect.'

More energy is generated on sunny days and less energy is generated on cloudy/rainy days, or when the photovoltaic array is shaded by obstructions (trees, lampposts, buildings, etc.).

On the other hand, thanks to the so-called 'cloud edge effect', the power generated by a solar panel may exceed its nameplate ('stated' or 'rated') power. The cloud edge effect is a common phenomenon upon which the overall solar irradiance increases as a result of the solar energy reflected by the edge of the clouds. This energy directly adds up to the solar power falling on the solar panel, thus increasing the power yield.

Important!

Solar photovoltaic panels generate DC electricity.

The generated electrical energy can be:

- Used right away by DC devices
- Stored in a battery for later use

- Converted to AC electricity and then either used by home appliances or exported to the utility grid (by grid-tied solar panel systems, which are beyond the scope of this book).

Solar photovoltaic modules

Photovoltaic solar modules consist of photovoltaic cells connected in series or parallel. In a solar module, the solar cells are typically connected in series to provide a higher voltage.

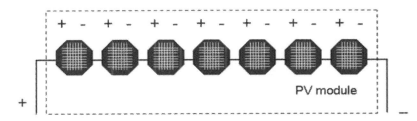

Solar modules are light and small enough to be installed on roofs, sometimes upon adverse conditions.

To get even higher voltage, you can connect the solar modules in series. A group of solar modules connected in series is called 'string':

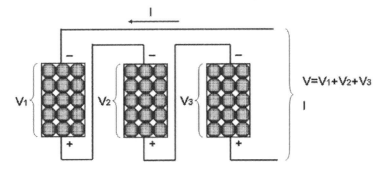

Efficiency is a primary parameter of solar modules. Efficiency shows what part of the solar energy fallen onto a solar module's surface is converted into electrical power.

Important!

Every solar module has its rated power or peak power, denoted in kWp.

The peak power of the module, however, is not the real power it can generate.

The real power of the module is always lower than its rated power, due to the following factors:

- Manufacturer power tolerance
- Dirt and dust
- Temperature
- Cable losses
- Inverter efficiency
- Shading.

Manufacturer power tolerance is the percentage within which the manufacturers guarantee that the real power output will be the same as the rated power output. Such a percentage is never 100%, a typical value is 95%.

Dirt and dust also cause losses when accumulated on the surface of a solar module. Dirt and dust particles can block the sunlight, thus reducing the power output. The content of dirt and dust in the air may vary with location and in an urban environment reaches its maximum. In regions with heavy rainfall, dirt losses tend to be zero.

Temperature is one of the most critical factors to be considered when designing a solar panel system. Temperature influences all the three main electrical parameters of a solar module – voltage, current, and power. When the weather gets warmer, the output voltage and generated solar power goes down, and when the weather gets colder, voltage and the generated power go up. The solar cell temperature, identifying whether the generated solar power goes up or down, is 25°C (77F), according to the Standard Test Conditions (STC).

A solar panel tends to have a negative temperature coefficient, which means that the solar panel performance declines when the temperature of a solar cell increases.

The solar panel's rated output power is defined under Standard Test Conditions (STC), as follows:

- 1,000 W/m^2 of sunlight
- 25°C of cell temperature (77F)
- Spectrum at air mass of 1.5.

Generally, for positive ambient temperatures, the temperature of a solar cell is about 15°C higher than the ambient temperature, as a result of the solar panel encapsulation.

Cable losses are inevitable in any solar panel system, especially in case of long cables which should be avoided whenever possible. A typical value of cable losses is 3% to 5%.

Inverter efficiency denotes how much of the input DC power is converted into AC power. The percentage is never 100, but values of inverter efficiency between 90% and 95% are commonly assumed in practice.

Shading must be avoided by all means, since even small shadows can severely reduce the performance of a solar module. A solar module consists of cells, and when shaded, every cell turns into a heat-dissipating device boosting dramatically the temperature of the module. This results not only in an unexpected reduction of the output voltage but also in shortening the lifecycle of the cells and modules. When mounted on the roof, solar modules might underperform due to the shading caused by trees, chimneys and other roof protrusions that are hard to eliminate.

Solar modules differ mostly in their:

- Type – monocrystalline, polycrystalline, thin-film.

- Power output (also known as 'power rating' or 'rated power') – between 10 Wp and 300 Wp.

- Output voltage – 12 V, 24 V, 48 V or 60 V.

- Size and weight – commonly 1.6 x 0.8 meters or 5.25 x 2.62 feet.

Solar module types

Monocrystalline modules

Monocrystalline modules are the most efficient, but also the most expensive ones. They come in blue or black color.

The fewer solar modules you need to produce a certain power, the higher their efficiency. When you do not have enough space on your roof, you should choose modules of higher efficiency.

Polycrystalline modules

Polycrystalline modules are slightly less efficient than monocrystalline ones and cost 30-50% less. Polycrystalline modules have a lifecycle of about 25 years. The practice has shown, however, that polycrystalline modules installed more than 25 years ago are still perfectly operational.

Polycrystalline modules are typically blue in color and can be easily distinguished by their multifaceted, kind-of-shimmering appearance.

Thin-film (amorphous) modules

Thin-film modules are the least expensive modules with the lowest efficiency – usually half of the efficiency of monocrystalline modules. Therefore, to generate the same power, you need twice as many thin-film modules as monocrystalline ones.

Thin-film modules have a dark surface usually colored in brown, grey or black. They are widely used in solar calculators.

Crystalline (mono- or poly-) solar modules are the most common type in residential and office solar panel systems. Crystalline modules come in a variety of size and shape. The rectangular shape is the most common. Photovoltaic modules have a lifespan of about 20-25 years, which means that about 80% of their rated power is guaranteed within such period.

Every solar panel has nominal ('rated') power measured in 'watts-peak' (Wp) or 'kilowatts-peak' (kW). Here is a comparison between solar panel efficiency according to the area needed to install a solar panel of nominal power 1 kWp:

PV cell material	Module efficiency	Area needed for 1 kWp
Monocrystalline silicon	13-16%	7 m² (75 sq.feet)
Polycrystalline silicon	12-14%	8 m² (86 sq.feet)
Amorphous silicon	6-7%	15 m² (161 sq.feet)

During the period between the first six months and one year of operation, thin-film modules produce about 10-15% higher output than initially intended. After that, within a period of another six months to one year, the power yield settles down to the rated value sustained over the remaining years of operation.

Electrical parameters of solar panels

Standard Test Conditions

The solar power falling per unit of area is called 'solar irradiance.' The energy produced depends directly on the sunlight falling onto the solar module's surface. The higher the irradiance, the more the solar-generated electricity.

The highest energy is generated by the direct radiation. The diffuse radiation results in a lower energy output.

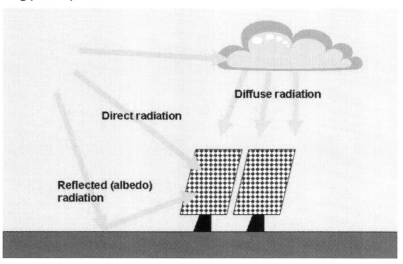

Important!

The maximum energy to be generated is defined upon the **Standard Test Conditions (STC)**.

The purpose of the Standard Test Conditions is to define stable conditions, thus enabling comparison of different solar cells and modules.

The STC are irradiance, cell temperature, and air mass:

- Irradiance E = 1000 W/m2 – the peak sun value

- Cell temperature T = 25°C (with a tolerance of ± 2°C) or 77° Fahrenheit

- Air mass (the atmosphere through which the sunlight must pass to reach the earth) AM =1.5.

The STC represent the varying conditions to which the solar modules are exposed once they are mounted. All these conditions affect the power output of a solar panel:

- Irradiance depends on the geographic location and season. The higher the irradiance, the higher the current.

- The higher the cell temperature, the lower the voltage.

- Air mass value depends on the geographic location, season and time.

Important!

To get the maximum sunlight and generated solar power, you should keep a solar panel cool and point it as directly to the sun as possible.

I-V curves

DC voltage is generated when a solar module is placed in the sun. DC current is generated when a solar module is placed in the sun, and the module terminals are connected to a load by closing the circuit.

When a solar module is not exposed to sunlight:

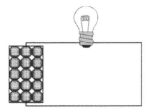

This is equivalent to an open circuit – no current is flowing:

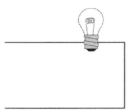

When a solar module is exposed to sunlight:

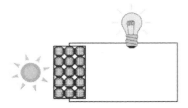

This is equivalent to a closed circuit:

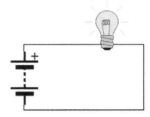

With voltage and current generated, power is also generated:

P [Watts] = V [Volts] * I [Amps]

Important!

The main electrical characteristics of a solar cell are its voltage and current.

The current generated by a solar cell depends on:

- The irradiance
- The cell area
- The voltage of the cell.

The voltage generated by a solar cell:

- Depends strongly on the cell material, and
- Practically does not depend on irradiance.

For silicon solar cells, the generated voltage is about 0.5-0.6 V per cell. Higher voltages are obtained by connecting solar cells in series:

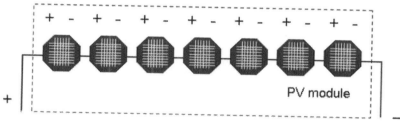

And higher currents are obtained by connecting solar cells in parallel:

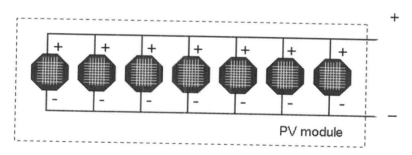

The relation between the voltage and the current in a solar cell or module is described graphically by the volt-ampere characteristic, also known as *I-V curve*. In other words, an I-V curve shows how the current in a solar cell/module is affected by a change in the voltage:

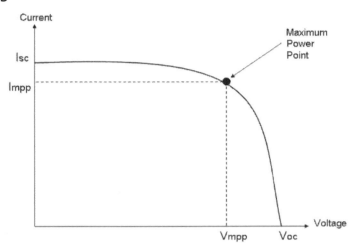

The most important points in a solar module's I-V curve are:

- The short-circuit current (Isc)

- The open-circuit voltage (Voc)

- The maximum power point voltage (Vmpp) and maximum power point current (Impp), defining the **Maximum Power Point (MPP)**.

The **short-circuit current (Isc)** is the maximum current that can flow if we short-circuit the positive and the negative end of a solar module exposed to sunlight. When we short-circuit a solar module, V=0 and I=max, since there is no load plugged into the circuit and the resistance R=0:

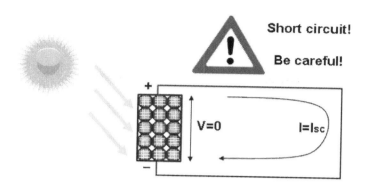

Short circuit!
Be careful!

Important!

The short-circuit current Isc cannot damage a solar module. Isc, however, is a hazard both for you and the cabling:

- For you, since it exceeds the maximum permissible current of the human body and can cause you a severe electrical shock!

- For the cabling, since if the maximum permissible current for the conductor is lower than **Isc** and the conducting wire in the cable gets more heat than it can dissipate, as a result of the extremely high energy of the moving electrons and due to the lack of any loads to render such energy on. This means a fire hazard.

Isc is not the operating current for a solar module. It is primarily used upon sizing a solar panel system.

The **open-circuit voltage (Voc)** is the maximum voltage that can occur between the positive and the negative terminal of a solar module exposed to sunlight.

Here, unlike short-circuiting, we have the opposite

67

case: V=max, the resistance R is very large and I=0, which means that the circuit is open and no current is flowing.

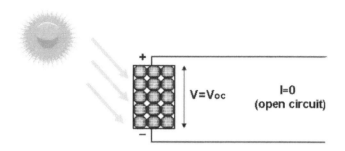

The open-circuit voltage **Voc** is the maximum possible voltage between the terminals of a solar cell or module.

Voc is not a real operating voltage for a solar module. It is primarily used upon sizing a solar panel system.

The **maximum power point (MPP)** is the point where the solar module produces the maximum power Pmax:

Pmax [kWp] = Vmpp * Impp

The basic electrical parameters of a solar module – **Vmpp, Impp, Isc**, and **Voc** – have different values for various photovoltaic materials – polycrystalline, monocrystalline, and thin film.

Impp and **Vmpp** are used for maximizing the power output of a solar panel system.

Since **Impp** and **Vmpp** correspond to the maximum generated power, and as everyone wants to have a system of best performance, solar panel systems are usually designed to operate at **Impp** and **Vmpp**. That is why **Impp** is often called 'operating current' and **Vmpp** is called 'operating voltage.'

The power generated at the **MPP** is the maximum possible power that can be produced by a solar module. It does not correspond to the maximum possible values of the voltage and current (**Voc** and **Isc**) but rather to the operating voltage and current (**Vmpp** and **Impp**). The power generated at the **MPP** is also called 'rated output power', and is measured in Watts (W).

Important!

Electrical parameters of solar panels can help you:

- Assess the solar panel system you can have
- Calculate the system cost
- Size the system correctly
- Select the right system components

The rated current, voltage, and power of a solar module can be found on its backside and in its specification sheet. They are really essential, as they provide you with the following key information:

- The maximum power informs how much power a solar module can produce and helps in calculating the energy output of a solar power system.

- The current indicates the maximum permissible value to consider when sizing the cables and the other system components (charge controller, inverter).

- The voltage shows whether a solar power system has enough potential to generate the current needed for the loads to be plugged into.

Power tolerance is the percentage within which the manufacturers guarantee that the real power output will be the same as the rated power output. The typical power tolerance falls between 0% and +3%. For example, if a solar module has a rated power output of 100 W and a power tolerance of 2%, you can expect that it will produce between 100 W and 102 W of power. The power tolerance is defined upon the STC (described above).

An essential parameter of any solar module is its **nominal 'peak' power** (already mentioned above) Wp. The peak power Wp is used:

- To estimate the maximum power output of an array of solar modules and match it to the components you will connect this array to – batteries, inverters, cables, etc.

- To calculate the expected solar-generated energy and match it to your seasonal energy consumption.

Impact of irradiance and temperature on solar panel performance

The sun provides not only sunlight but also heat. Sunlight is the solar panel's friend, while heat is the solar panel's enemy. The more intensive the sun, the more sunlight provided and the more efficient a solar panel. You have more current and produce more power, which is good. The more intensive the sun, however, the more heated the solar panels, and hence the lower the voltage generated. Unfortunately, the voltage decrease is much higher than the current increase. This means a lower energy output, which is not good.

To avoid surprises, you have to consider the effect of the heat. Since we should not let the heat spoil our work, we need to take into account any possible voltage decreases due to heat, so that the performance of our system would not be affected.

Important!

As temperature increases, the generated voltage decreases.

You have to consider the effect of the temperature on the system voltage and performance, since solar arrays and panels, being outdoor-mounted, are exposed to really extreme conditions both in summer and winter.

Although voltage is affected by temperature, your system should be designed so that its power output is not affected by any seasonal temperature changes.

Solar modules mounting

Mounting structures hold solar modules and arrays in the proper place. A solar panel mounting structure is supposed to bear various kinds of loads – wind, snow, mechanical pressure, thermal influence, as well as its own gravity.

The solar array mounting and the related cabling should be reliable yet simple to enable easy replacement of the panels, without the need of dismantling the whole solar system. Racking components hold the solar array to the spot where the solar modules are mounted.

There are various ways of mounting a solar array. The most common option is to mount it on a roof. Often a solar array can also be mounted on the top of a pole rack – as is the case of solar lamps in parks.

Important!

The solar array mounting type should be selected by carefully considering:

- The orientation towards the sun
- The site shading
- The weather at the location
- The roof material and bearing capacity (in case of roof mounting)
- The type and condition of the soil (in case of ground-mounting).

In off-grid solar power systems, often panels are not mounted on roofs but rather on a ground-mounted construction or pillar.

Connecting photovoltaic modules

In order to produce as much power and energy as possible, photovoltaic modules are connected in series or in parallel. To avoid loss of power in a solar panel system, you should only connect solar modules of the same type. Connecting solar modules in series means joining the positive terminal (+) of a module to the negative terminal (-) of the next module, joining the negative terminal of that module to the positive terminal of the next module, and so on:

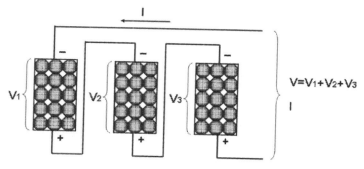

$$V = V_1 + V_2 + V_3$$

A set of solar modules, connected in series, is known as 'string.' Solar modules are connected in series to obtain a higher output voltage. The maximum system voltage, however, must not be exceeded.

For modules, connected in series, the total power is calculated as follows:

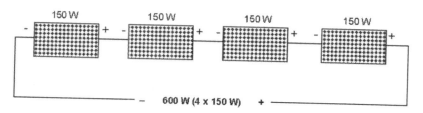

Total connected power = 150W + 150W + 150W + + 150W = = **600W**

Solar modules are connected in parallel in the following way:

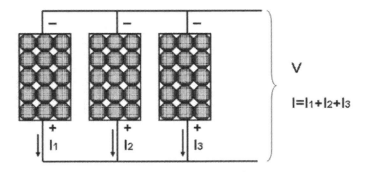

In parallel connection, the positive (+) terminals are joined together and the negative (-) terminals, in turn, are joined together. Solar modules are connected in parallel to obtain a higher output current. For solar modules connected in parallel, the total power is calculated as follows:

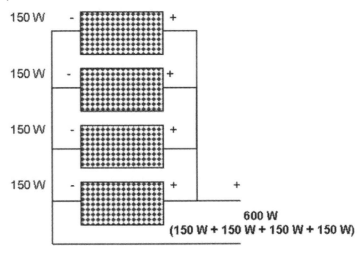

Total connected power = 150W + 150W + 150W + + 150W = = **600W**

For off-grid systems, the parallel connection of modules is more typical.

Examples of connected solar modules

The following general principles are valid for series and parallel connection:

- For series connection, the current flowing through each load is the same, while the total voltage on all the loads is the sum of the individual voltages on the loads.

- For parallel connection, the voltage on each load is the same, while the total current is the sum of the individual currents through the loads.

- Whether connected in series or parallel, solar modules produce the same power.

Mixed connection is also possible, with the same principles being valid.

Example 1) Series connection of two modules

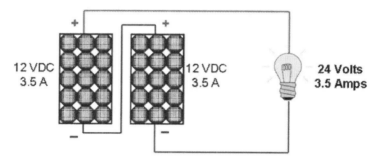

Total voltage = 12V + 12V = 24 Volts

Total current = 3.5 Amps

Total power output = 24 Volts * 3.5 Amps = 84 Watts

Example 2) Parallel connection of two modules

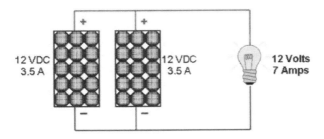

Total voltage = 12 Volts

Total current = 3.5A + 3.5A = 7 Amps

Total power output = 12 Volts * 7 Amps = 84 Watts

Example 3) Series connection of four modules

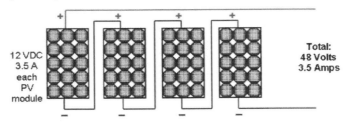

Total voltage = 12V + 12V + 12V + 12V = 48 Volts

Total current = 3.5 Amps

Total power output = 48 Volts * 3.5 Amps = 168 Watts

Example 4) Parallel connection of four modules

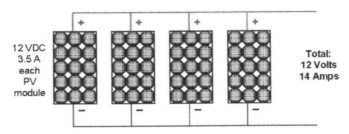

Total voltage = 12 Volts

Total current = 3.5A + 3.5A + 3.5A + 3.5A = 14 Amps

Total power output = 12 Volts * 14 Amps = 168 Watts

Example 5) Mixed connection of four modules

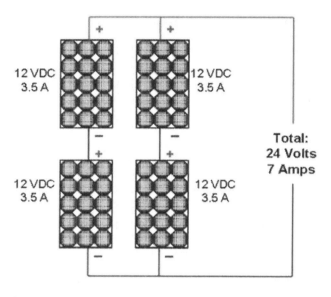

Total voltage = 12V + 12V = 24 Volts

Total current = 3.5A + 3.5A = 7 Amps

Total power output = 24 Volts * 7 Amps = 168 Watts

Solar radiation and site selection

To make solar systems work well for you, you need to know the basics of solar radiation. Also, you should know how much sunlight your system will receive and how specific changes can affect the system performance.

Furthermore, it is essential not only to design your off-grid solar system correctly but also to mount the panels at the right place. You should also know where to find the right tools that can help you evaluate your system and take the right decision.

Solar radiation

Total radiation at a site is a sum of direct and diffuse solar radiation.

The direct radiation comes from the sun's direction and can be overshadowed by objects and buildings. The diffuse radiation does not come from a specific direction, since it is scattered from the sky.

The direct radiation is more significant for the solar array's operation than the diffuse radiation, since it affects better the ability of the array to convert the solar energy into electricity.

On a bright and sunny day, the direct radiation is predominant over the diffuse one. On a cloudy day, the diffuse radiation from the sun is predominant, while the direct radiation can be even missing.

Unfortunately, cloudy and foggy days are of less use for any solar panel system, since the diffuse radiation plays a relatively insignificant part in the solar-generated output.

For proper dimensioning of solar panel systems, it is vital to know the path of the sun. This path is used for calculating the solar irradiance.

'Irradiance' is the intensity of the solar radiation falling onto the earth's surface. The irradiance can also be regarded as the solar power falling onto a specific area. It is measured in W/m^2 or kW/m^2.

Important!

The irradiance falling over a solar module depends on:

- The location of the solar module
- The position of the module towards the sun
- The season
- The weather.

Here is how the electrical current produced by a solar module depends directly on the irradiance:

- The current increases when the irradiance increases.
- The current decreases when the irradiance decreases.

Voltage is less dependent on the irradiance. A solar module is capable of generating voltage even during a cloudy day.

Since the power output of a solar panel depends both on the voltage and current, the electrical power produced by a solar system depends directly on the irradiance, and also on how long the sunlight has been falling on the solar panel surface.

Therefore, the fewer the sunny days, the less electricity produced. In case of insufficient sunlight, battery systems guarantee that the electricity in the house should be always on.

Another essential feature is irradiation. Irradiation is the irradiance measured for a day, a month or a year. When designing a solar panel system, the period where the **daily** value of irradiance equals 1,000 W/m^2 is the most important. This period is called 'Peak Sun Hours' (**PSH**).

The PSH for a particular area defines how long on a sunny, cloudless day a solar array operates at its maximum power output. For a particular location, PSH is measured in hours per day.

Important!

- **Peak Sun Hours (PSH) is a measure of the available daily solar resource.**
- **PSH depends on the specific location.**
- **PSH is used to calculate the energy output of a solar panel system.**

Do not mistake Peak Sun Hours for available sunny hours. For example, six sunny hours during a bright sunny day might only translate into four Peak Sun Hours for that day!

Every solar system should be designed so that to use the energy of the sun in the most efficient manner. The point is to provide your solar panels with full access to as much sunlight as possible. To design a photovoltaic system, you need to find how much sunlight the system can receive when installed at a specific location. For example, if the PSH of your location is 3 hours, this means that your solar system will receive 3 hours of 'full' sun as an average.

Important!

To estimate the power output of a solar module or array, you need the following information:

- The value of the solar irradiation (also known as 'solar insolation' or 'Peak Sun Hours', **PSH**) for the specific longitude and latitude of the location.

- The orientation of the solar module/array – to the South (if the point is located in the Northern hemisphere), East, West or North (if the point is located in the Southern hemisphere – e.g., South Africa, Australia or New Zealand).

- The azimuth and tilt angle.

- The nominal 'peak' power of the solar module/array **Wp**.

- The System Loss Factor **SLF**.

The daily energy output of a solar module can be calculated by:

E = Wp * PSH * SLF

Where:

- E is the solar module's daily energy output in Watt-hours (Wh)

- Wp is the solar panel nominal 'peak' power

- PSH is the Peak Sun Hours per day

- SLF is the System Loss Factor.

The System Loss Factor (SLF) is a dimensionless value usually falling between **0.5** and **0.7** for off-grid solar power systems. It accounts for all system losses between the solar panels and the load. The higher the SLF value, the better, as more solar power is transferred from solar panels to the load with minor losses. The ideal case is when SLF=1. In this case, all the solar-generated electricity is delivered to the loads without any losses. In the real world, however, achieving SLF=1 is impossible, since in every solar system a part of the solar generated electricity is lost in the system components (solar panels, batteries, charge controller, inverter), cables, and also as a result of non-optimal positioning of the solar array. To get all the possible losses included in the SLF in details, please refer to our book *"The Ultimate Solar Power Design Guide: Less Theory More Practice"*.

Solar maps

PSH is obtained by solar maps. Solar maps picture the values of the available solar resource for any location in the world. They help you estimate how much energy your solar panel system can produce daily, monthly or annually. Solar maps can be used either in a graphical form or as a software database. The latter is much more precise and provides more opportunities when designing a solar panel system.

Several types of maps about PSH are available online:

- Maps showing the irradiation (in kWh/m^2) on a horizontal plane. Such maps are of least use, since they consider neither the azimuth nor the tilt angle.

- Maps showing the irradiation (in kWh/m^2) on a horizontal plane mounted at a specific azimuth.

- Maps showing the irradiation (in kWh/m^2) on a horizontal plane mounted at a specific azimuth and tilted at a particular angle. Such maps provide the most detailed information and are used for obtaining the most meaningful results.

Daily values of solar irradiation are represented as averaged for a day, month or year. Therefore, it is possible to calculate the power output of a solar module/array for a day, month or year.

You can find solar radiation databases at:

http://photovoltaic-software.com/solar-radiation-database.php

https://eosweb.larc.nasa.gov/cgi-bin/sse/grid.cgi?

https://asdc-arcgis.larc.nasa.gov/sse/

Solar maps are divided into:

- **Solar maps depicting the total solar irradiation on a horizontal service.** Such maps are of the least use, since any solar array produces the least power when placed in a horizontal position.

- **Solar maps depicting the total solar irradiation on a surface mounted at an angle equal to 'latitude,' 'latitude+15°' or 'latitude−15°.'** Such solar maps are more useful.

- **Solar maps depicting the total solar irradiation on a surface mounted at a specific angle optimal for a given location.** Such solar maps are the most useful ones due to their accuracy. Often they are generated by software databases and can be found online for free.

Important!

A precise and reliable online source for solar data is **NASA Surface Meteorology and Solar Energy Data Set** (see Appendix B):

https://eosweb.larc.nasa.gov/cgi-bin/sse/grid.cgi?

By entering the latitude and longitude of a location, you get the daily PSH, averaged both annually and monthly, for various tilt angles, including also the optimal one:

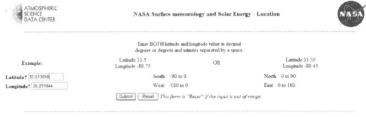

NASA solar database, courtesy of NASA

85

Both NASA database and graphical solar maps are used to determine the average annual PSH when calculating the annual electricity production of a solar electric system. Graphical solar maps, however, are usually not suitable to determine the PSH for the worst month, used for evaluating off-grid systems. The worst month PSH should be obtained from the NASA's database, which is much more precise.

Azimuth and tilt angles

The two critical parameters related to the solar module positioning are azimuth and tilt:

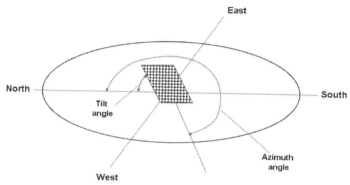

Azimuth (orientation) is the angle between the direction perpendicular to the array's surface and North (if you live in the Northern hemisphere – USA, Canada, and Europe) or South (if you live in the Southern hemisphere – Australia, New Zealand, South Africa).

Tilt (elevation) is the angle measured between a mounted solar module and a horizontal ground surface:

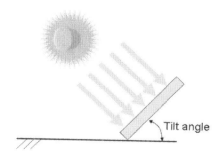

As a rule, solar arrays are recommended to install on roofs facing South (for USA, Canada, and Europe) or North (for Australia, New Zealand, South Africa).

Installation of panels on roofs facing North (or South – if you live in the Southern hemisphere) is **NOT** recommended.

Azimuth

When designing a solar panel system, it is vital to mount the solar array correctly, so that it would receive as much sunlight as possible.

If you want to achieve the maximum performance, you should keep your solar array perpendicular to the sun as long as possible, since the generated current directly depends on the irradiance.

Indeed, the sun moves all day long, so it is not easy to move the array all the time to keep it perpendicular to the sun. It is possible, however, to find an average position for mounting the array. Such a position is reached by directing the array towards the equator at an angle approximately equal to the latitude of your location.

Stationary mounted arrays should be oriented to face the average position of the sun. Usually, such arrays have fixed tilts, of a tilt ranging from 0° to 50°.

Important!

A solar module produces the maximum power when facing the sun.

A solar module facing 30° away from South (or North) will lose 10-15% of its power output.

If a solar module faces 60° away from South (or North), the power loss is 20-30%.

Important!

A rule of a thumb for determining the azimuth:

- If a solar power system is located in the Northern hemisphere, the solar modules should point as close to the South as possible.

- If a solar power system is located in the Southern hemisphere, the solar modules should point as close to the North as possible.

You need to find the best azimuth for your solar array for summer-only, winter-only or annual energy production. What is more, the higher the latitude (that is, the further you live from the equator), the more significant the difference between the azimuth required to achieve the best performance in summer and winter.

Tilt

Here is how the tilt affects the performance of a solar module:

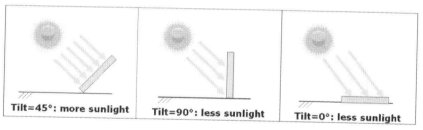

Tilt=45°: more sunlight   Tilt=90°: less sunlight   Tilt=0°: less sunlight

For off-grid solar systems based on fixed-mounted solar modules, the tilt of the solar modules should be determined for optimal system performance at the scarcest sunlight. It is calculated that if the electric power output is enough upon the scarcest sunlight, the selected tilt is suitable during the rest of the year as well.

Important!

For most geographical locations, the optimum tilt should fall somewhere between '**latitude**' and '**latitude−15°**.'

Source:

Pop MSE, Lacho,Dimi Avram MSE. 2015. The Ultimate Solar Power Design Guide: Less Theory More Practice, Kindle Edition. Digital Publishing Ltd.

Batteries

Introduction

Batteries are devices capable of producing and storing DC electricity. In solar panel systems, batteries are used to replace the photovoltaic generator:

- At night,
- During cloudy weather, or
- While the solar array is disconnected for repair and maintenance works.

A battery cell is a container usually filled with diluted sulfuric acid used as an electrolyte, with two electrodes (of positive and negative polarity) immersed into it. The electrodes are made of grid-shaped lead plates. Such battery cells are called 'wet cells.'

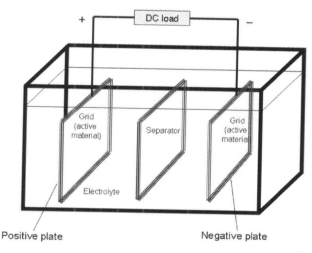

The two electrodes – one positive and one negative – are made of an active material that reacts chemically with the electrolyte, which results in generating current as a flow of electrons moving from the negative to the positive pole.

Cells are placed inside a shared housing and are internally connected in series. For example, six 2V-cells should be connected in series to obtain a battery block with a total voltage of 12V. Battery cells are connected to form a battery. Batteries, in turn, are connected to form a battery bank.

Capacity is the most significant battery parameter. It shows the current a battery can provide for a particular number of hours. A battery of 100Ah capacity can provide either 100A within 1 hour or 10A within 10 hours.

The optimal charge/discharge current for a battery of 100Ah capacity should be 10A. Such a battery can deliver this charge/discharge current for 10 hours.

An important factor affecting capacity is storage temperature. The lower the temperature, the less capacity provided. In general, low temperatures decrease the battery capacity, while high temperatures reduce the available battery lifecycles. One cycle encompasses one charge and subsequent discharge of a battery. Working temperatures between 20°C (68F) and 30°C (86F) are considered an optimal tradeoff.

After DC electricity is stored in a battery, it can later be rendered as DC voltage.

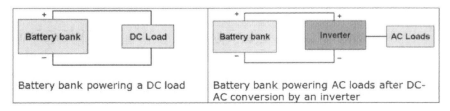

| Battery bank powering a DC load | Battery bank powering AC loads after DC-AC conversion by an inverter |

Standard battery voltages are 6V, 12V, 24V, and 48V.

The higher the capacity, the longer the period of voltage delivery. After some time, a battery is no longer capable of delivering its rated voltage. At this point, the battery is said to be in a state of 'discharge.' To regain its capability of delivering the rated voltage, the battery needs to be recharged.

Important!

The shorter the discharge/charge period, the shorter the battery lifespan.

For optimal battery lifespan, the charge/discharge current should not exceed 1/10 of the battery capacity. Thus, for a battery of capacity 55Ah, the charging current should not exceed 5.5 Amps. Every manufacturer provides thorough instructions about the optimal charging and discharging of their batteries and it is recommended to stick to these guidelines.

In a solar panel system, the battery bank is recharged by the solar array:

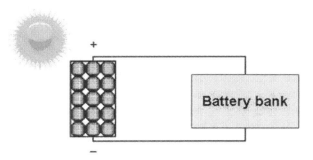

Batteries used in solar systems are rechargeable. Each battery has its lifespan, which means that it cannot be discharged and recharged illimitably.

Important!

Vehicle batteries are not designed for frequent and deep discharges. For this reason, vehicle batteries are not recommended for solar power systems.

Depth of Discharge (DoD) is the percentage down to which the capacity can be reduced during discharging (100%=empty battery, 0%=full battery). The lower the DoD, the longer the battery lifespan. A battery regularly discharged down to 80% of its capacity will have a shorter lifespan than a battery regularly discharged to 50% of its capacity. For deep-cycle batteries, DoD is about 80%, and for liquid electrolyte batteries, DoD is 50%. The most practical DoD value to use when sizing a lead-acid battery bank is 50%.

Days of Autonomy (DoA, holdover) is the number of days a battery can support a load, without any need to be recharged by the solar array. The more the days of autonomy, the more expensive a battery. DoA depends on whether the off-grid solar power system is standalone or hybrid. **The typical value of DoA is 2-3 days** for hybrid off-grid systems and **3-7 days** for standalone (photovoltaic-only) systems.

The **end of battery life** is reached when a battery is capable of maintaining about 70-80% of its original capacity.

In solar panel systems, batteries must also be prevented from **overcharging** and **overdischarging**. Overcharging can lead either to hazardous condition or to shortening the battery lifespan. The same is valid for overdischarging. In an off-grid solar electric system, it is the charge controller that prevents the battery bank from overdischarging and overcharging.

Batteries in off-grid systems

Since in most off-grid solar panel systems, the solar-generated electricity is seldom entirely consumed right away, it is stored by batteries. Batteries in off-grid systems have three essential features:

- Autonomy (the main advantage of off-grid systems) – through proper selection of the battery parameters (capacity and depth of discharge) the system power output is matched to the energy needs of a household, without any dependence on the utility grid.

- Provision of a stable DC voltage enabling the inverter or the loads to work properly and not to get damaged.

- Ability to provide high currents. Such high currents are needed upon starting motors when surges (abrupt and high increases in current, known as 'spikes') are likely to occur.

The cost of batteries alone can reach between 25% and 50% of the total cost of an off-grid system. Batteries of longer lifespan cost more, but are less expensive to maintain. Car batteries are not suitable for solar panel systems, as they are not designed for deep discharges.

In off-grid systems, the following types of batteries are used:

- Lead-acid batteries – flooded and sealed (VRLA)

- Alkaline batteries – nickel-cadmium (for small portable appliances or daily loads) and nickel-iron

- Lithium-Ion batteries – used in electric vehicles

and small airplanes, cell phones and other portable electronic devices; currently they are trying to find their place in solar power systems.

Lead-acid batteries

Lead-acid batteries comprise multiple individual cells of nominal voltage 2V each. Lead-acid batteries contain 'wet cells' where the electrolyte is in a liquid state.

Important!

Lead-acid batteries are the best choice for solar panel systems. They are said to offer '**the best value for the price**' and have the following benefits:

- Low cost
- Robust design
- High depth of discharge.

Lead-acid batteries are flooded and valve-regulated ones.

Flooded lead-acid batteries have removable caps and require regular maintaining activities – checking the level of electrolyte and adding distilled water when the level falls below a required minimum. The caps should be kept opened during the process of recharging to free the hydrogen gas released. For stationary off-grid systems, flooded lead-acid batteries are recommended, due to their:

- Longer lifespan
- High capacity
- High reliability
- High depth of discharge
- High performance (higher than gel batteries)
- Lower cost.

Flooded batteries have their drawbacks, as they require:

- A separate, safe and well-ventilated room to install
- Regular maintenance activities
- Serious attention and high commitment by the user.

Valve-regulated lead-acid (VRLA) batteries are also known as 'sealed lead-acid batteries.' During recharging, the pressure of the hydrogen released pushes the valves up and releases the gas.

VRLA batteries have the following advantages:

- Do not require maintenance by the user regarding regular level checking and adding water.
- Produce less hydrogen gas during the recharge process.
- Are suitable for locating in a limited room space.

VRLA batteries, however, are not widely used and still are not preferred to flooded lead-acid batteries, due to the following reasons:

- More expensive
- Shorter lifespan
- The charging voltage should not be exceeded.

Alkaline batteries

AlkaGel (sealed) batteries are suitable for **mobile** applications of **small, low-current** off-grid solar systems.

Advantages of gel batteries:

- Easy for handling and transportation
- Safety (no acid, no gassing during charging)
- Maintenance-free (no distilled water to add).

Gel batteries have some disadvantages:

- They have to be replaced every few years
- Overcharging reduces battery capacity
- High cost – far from unaffordable, but anyway much more expensive than flooded batteries.

Nickel-cadmium batteries are an excellent choice for mobile applications, as they:

- Have a high number of discharge cycles
- Are less affected by temperature changes than lead-acid batteries
- Are less dangerous
- Are maintenance-free
- Are easy for transportation.

For residential applications, however, nickel-cadmium batteries are not advantageous, due to their high price and tough recycling. But if you can afford a nickel-cadmium battery for your home, it is the best choice.

RV (Recreational Vehicle) or Marine batteries are another battery type used in small solar power systems for boats or campers. Despite being deep-cycle batteries, they differ from typical deep-cycle batteries, as follows:

- They are smaller in size and less expensive.

- They are usually either AGM or gel-ones (i.e., maintenance free).

- Their voltage is usually 6V – to get a voltage of 12V, you should connect two batteries in series – as is the case with golf cart batteries).

- Withstand to fewer (3 to 4 times) charge/discharge cycles than residential deep-cycle batteries.

- Can be used more successfully to start an engine despite the lower performance compared to the typical vehicle batteries.

RV/Marine batteries are a compromise between vehicle batteries (also known as 'cranking' batteries) and deep-cycle batteries (typically used in residential solar power systems). RV/Marine batteries are used in small photovoltaic systems where size is often as important as performance.

Li-Ion batteries

Despite the infamous Sanyo laptop recall during 2006/2007 provoked by lithium-ion batteries for laptop power supply causing a fire hazard, this type of battery has already found its place in solar power systems. These batteries are mostly used for storage in non-solar or other non-renewable applications.

Advantages of Li-Ion batteries

- Low weight – about 30% of the size of a lead-acid battery of the same capacity.

- Lower space for housing – about 50% of the space needed for housing a lead-acid battery of the same capacity.

- Faster charging due to the higher recommended charge current for ensuring a maximum battery lifespan – about 30% of battery capacity compared to recommended 10% of the battery capacity for lead-acid batteries.

- Higher maximum discharge current.

- Recommended maximum discharge depth 70-80%, compared to 50% for lead-acid batteries.

- Twice longer battery lifespan under recommended usage conditions, which translates into about 10 years for Li-Ion batteries, compared to 5 years for lead-acid batteries.

- Provide more capacity in low-temperature conditions.

- Fully maintenance-free.

Disadvantages of Li-Ion batteries

- Higher price, although the high initial investment in a long run makes these batteries the lowest cost option due to their longer lifespan and higher number of charging cycles.

- A battery management system is a must.

- The battery voltage operating window might be incompatible with the input operating window of the inverter, as most inverters are manufactured to operate with lead-acid batteries.

Important!

Although the higher price of lithium-ion batteries could be justified by the higher lifespan, the requirement for a battery management system is a significant hurdle. The Battery Management System (BMS) prevents the battery from excessive charging and discharging by tracking and managing the voltage and temperature of an individual cell in a lithium-ion battery. Last but not least, through communication with the charge controller, the BMS saves Li-Ion batteries from overheating and possible fire hazard. Most lithium-ion batteries are provided with a BMS. However, not all solar charge controllers are BMS-compatible. The problem can come from not establishing a mutually understandable communication line between the controller and BMS due to lack of common communication platform or protocol. Therefore, you should perform a due diligence before buying a charge controller, even when it is stated that a particular charge controller model is designed to work with Li-Ion batteries.

Li-Ion batteries in summary:

- Today the application of lithium-ion batteries for solar power is viable in cases where lack of available space and low weight are more important than price.

- In spite of their higher price, lithium-based batteries can turn out to be a lower cost solution in the long run for solar power systems, compared to lead-acid batteries.

Connecting batteries

Series connection

For batteries connected in series, the total voltage is a sum of the individual battery voltages, while the total capacity is the capacity of an individual battery:

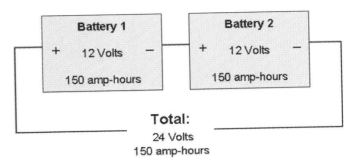

Important!

If you connect batteries of **different capacity** in series, the total capacity obtained is equal to **the lowest capacity** in the string.

Here is an example:

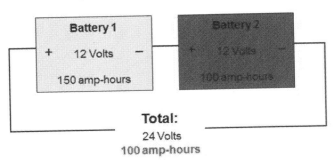

Important!

Avoid connecting in series a set of batteries of different capacity!

Parallel connection

For batteries connected in parallel, the total capacity is a sum of the individual capacities of the connected batteries, while voltage is the voltage of an individual battery:

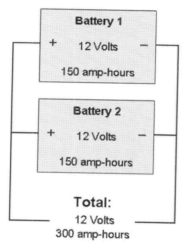

Important!

Recommendation: Connect identical batteries in series!

If you need a higher capacity, you should connect maximum 2 or 3 strings in parallel. Otherwise, the defective battery in a parallel string will start acting as a load to the adjacent 'good' string connected in parallel. Thus, the overall battery bank capacity will be reduced.

A string comprised of many low-voltage high-capacity batteries connected in series is a better option for getting the desired capacity than a string composed of one or two high-voltage batteries connected in parallel.

Furthermore, it is difficult to maintain an even charging and discharging per string due to the small difference between the voltages of each string. This is another reason for limiting the parallel strings up to 2 or 3.

The picture below shows a battery bank comprising a mixed connection of batteries to increase both overall voltage and capacity:

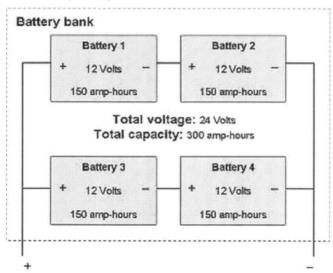

Source:

Pop MSE, Lacho, Dimi Avram MSE. 2016. The New Simple and Practical Solar Component Guide, Kindle Edition. Digital Publishing Ltd.

Charge controllers

The primary function of a charge controller is to prevent the battery bank from overcharging and overdischarging. Therefore, charge controllers (also known as 'charge regulators' or 'battery chargers') are used for proper maintenance of the battery bank.

Important!

Charge controllers control DC loads only.
AC loads are to be controlled by the inverter.

One of the most common problems of batteries is that they cannot be discharged excessively or recharged too often. A charge controller controls the charge by managing the battery voltage and current.

The charge controller is intended to protect the battery and extend its lifespan while keeping the system efficiency.

Important!

The primary functions of the charge controller are:

- Protecting the battery from overcharging by limiting the charging voltage.

- Protecting the battery from deep and/or unwanted discharging. The charge controller automatically disconnects the loads from the battery when the battery voltage falls below a certain Depth of Discharge value.

- Preventing the reverse current to flow from the battery bank through the solar modules at night, thus protecting both the panels and the batteries from damaging.

- Providing information about the battery state of charge.

- Squeezing more power from the sun by tracking and matching the optimal power operating point of the solar array (valid for MPPT controllers only)

Other essential functions of charge controllers:

- Switching between the various recharging modes, depending on the specific battery type, thus maintaining the battery in a good condition and increasing its lifespan and capacity.

- Protection against overloading and short-circuiting.

- Temperature compensation – application of a voltage-controlled single recharging mode matched to the battery temperature.

- Protection against overvoltage.

- Indicator lights for battery charging level, charging current, charging voltage, remaining time to get to the fully charged state, etc.

- Further charging of the liquid electrolyte cells.

For higher currents, more than one controller should be used. In such a case, the solar array should be divided into sub-arrays, and each sub-array should be connected to its controller. All of them, however, can use the same battery bank. For more information on this topic, please, refer to the section 'Scaling an off-grid solar power system' at the end of this chapter.

Charge controller types

PWM (Pulse Width Modulation) and MPPT (Maximum Power Point Tracking) are the main types of charge controllers.

PWM controllers are less expensive than MPPT ones. A PWM controller prevents the batteries from overcharging and overdischarging in a way extending the battery lifespan slightly more than an MPPT charge controller does, at the expense of lower (20% on average) efficiency.

A PWM controller achieves such extension of battery lifespan as a result of applying a complex switching algorithm and pulse width modulation combined with pulse charging. The pulse charging prevents the lead plates of the battery from sulfation.

Important!

A PWM charge controller is the best option, if:

- The solar array comprises just a couple of modules, and the installed solar power does not exceed 1.5-2kWp (1,500-2,000Wp).

- The nominal voltage of a single module is close to the battery voltage.

- The solar system operates in a warm climate.

The average efficiency of a PWM controller is about (75-80)% compared to the (92-95)% efficiency of an MPPT controller.

An MPPT controller prevents batteries from overcharging and overdischarging while transferring more power (up to 20-30% more) from solar modules to either the battery bank or the loads. It does not extend the battery lifespan as efficiently as a PWM controller does.

Some MPPT controllers are provided with a 'voltage step-down' feature. This feature allows a solar array of a higher voltage to be connected to a lower voltage battery bank – for example, a 48 V array to a 24 V battery bank. Without a step-down feature provided, the total voltage of the array must be equal to the battery bank voltage.

Important!

The step-down feature provides the following benefits:

- Higher solar array voltages – they allow the use of cables of smaller cross sections between the array and the controller, thus reducing the cabling costs. Therefore, more solar modules can be connected in series and fewer modules in parallel, compared to the case with no step-down feature available.

- An opportunity for solar array expansion without any need to increase the size of the cabling.

MPPT charge controllers are designed to track, fast and with maximum accuracy, the maximum power point of the solar panel/array under variable atmospheric conditions. Depending on how successfully an MPPT controller can do this, it can show up to 10% better performance by squeezing more power than its inferior MPPT counterpart.

During cloudy weather and continuously changing sunlight intensity, an ultra-fast MPPT controller can achieve up to 30% more efficiency than a PWM controller and up to 10% efficiency than a slower MPPT controller.

PWM and MPPT controllers – a detailed comparison

PWM controllers Pros:

- Prevent batteries from overcharging and overdischarging by extending the battery lifespan slightly more efficiently than MPPT controllers.

- Less expensive than MPPT controllers – the price starts from $25 to $250, depending on the power required.

- Most suitable for systems with installed solar power up to 1.5kWp (1,500Wp). For installed solar power higher than 1.5kWp, an MPPT controller is recommended. In such a case, up to 30% more energy yield will compensate for the MPPT controller's higher price in the long run. This is especially valid for colder climates.

PWM controllers Cons:

- Not optimized to ensure optimal load to solar panels, which translates into up to 20% average loss of solar-generated electricity. To compensate such efficiency losses, you should add more panels to your solar array.

- The system voltage must exactly match the solar panel nominal voltage, i.e., you have to use solar panels of 12 V nominal voltage, charge controller of 12 V nominal voltage, and a battery bank of 12 V nominal voltage.

- For this reason, low-cost 60-cell modules from the grid market might not operate well with PWM controllers. Due to their lower maximum power tracking voltage (~15V instead of ~18V for 72 cell modules) at higher ambient temperatures, 60-cell modules might not be capable of charging the battery bank.

- Solar array voltage should be slightly higher than the battery bank voltage. Otherwise, you will lose some potentially generated power from the solar array, since a PWM controller does not convert this excess voltage into current as an MPPT controller would do to increase the power output. A PWM controller just reduces the voltage of the solar array to a lower, yet high enough, value to allow the current from the array to charge the battery bank. This is the primary cause of the PWM controller inefficiency in lower temperatures. The lower the temperature, the higher the solar array voltage and the more energy lost for adjusting such a higher voltage to a lower one.

- PWM controllers might create an audio- or RF noise due to the Pulse Width Modulation used.

- Not very suitable for scaling up your system. You can get PWM controllers rated at up to 120 Amps. The most common ratings, however, are up to 60A.

MPPT controllers Pros:
- Prevent batteries from overcharging and overdischarging.

- 'Squeeze' more power (up to 20% on average) by ensuring an optimal load to the solar array. The colder the climate, the higher the squeezed power, as the solar array's operating voltage reaches its maximum.

- Reach maximum efficiency in cold climates (up to 30%-40% increase).

- The nominal voltage of the solar panels can be higher than the system voltage, i.e., solar panels of 48 V nominal voltage can be connected to a charge controller charging a 12V-battery bank.

- You can get MPPT controllers rated at up to 200A. However, the most common ratings are up to 80 Amps.

- For an MPPT controller, the warranty is typically longer than for a PWM one.

- An MPPT controller provides a better opportunity for system growth than a PWM controller.

- Some MPPT controllers (offered by MorningStar Corporation) may operate with an oversized solar array, of installed power that is several times higher than the maximum input power supported by the controller. These controllers just limit the output charging current to its maximum supported value without shutting down the controller, as some of their counterparts do when overloaded. This can be useful when you want to use cost-effective high-wattage panels rather than low-wattage ones or you wish to harvest more energy on average. In case of an oversized solar array, due to the limitation in the charging current of the solar array, the energy gain during periods of reduced sunlight can surpass the energy losses during sunny periods.

- It is known that 60-cell solar panels are not suited to operate well with PWM controllers.

However, lower-cost higher voltage 60-cells solar panels intended for grid-connected systems can be used with MPPT controllers in off-grid systems.

MPPT controllers Cons:

- More expensive – a good one starts from $500 to $x1,000.

- In hot climates, MPPT controllers lose their main advantage – the high efficiency. In a hot climate, the efficiency of an MPPT controller decreases to only 10% higher (on average) compared to PWM controllers, while in cold climates solar energy squeezed from an MPPT controller might increase up to 40%, compared to PWM controllers.

Scaling an off-grid solar power system

Scaling a solar power system means connecting two or more charge controllers to a battery bank in parallel, thus increasing the solar power generated. Both residential and mobile off-grid systems can be scaled.

You can scale up a solar power system by adding a second charge controller to charge the battery bank, with its solar panels, in parallel with the existing ones.

Here is a diagram for paralleling two PWM charge controllers.

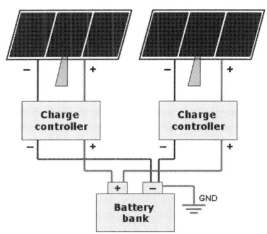

System scaling can be useful in the following cases:

- You do not have enough space on the roof, and you want to install a second solar array oriented in a different, sub-optimal direction.

- You want to avoid a power degradation of the existing solar array in case of partially shaded solar panels during the small part of the day. In such a case, you should form a separate solar

115

array from those shaded solar panels and connect it to a separate charge controller.

- You want to increase the charging power of your mobile solar system when camping. You can use the existing mobile solar array with a standalone primary charge controller and add on demand a portable solar array tilted at the optimal angle towards the sun, equipped with its charge controller connected in parallel to the primary controller.

For optimal performance, it is recommended every charge controller to be connected to its solar panels and to use the same battery charging program or settings.

You can connect the DC loads to the DC load terminals of each charge controller, provided that the current of the DC loads does not exceed the maximum current of each controller.

You can use the same layout to parallel MPPT controllers. In this case, you should follow the similar guidelines:

- Each MPPT controller should be connected to its solar array sized according to the pertaining controller specifications.

- The battery bank should be sized to support the charging currents.

- The MPPT voltage regulation should be accurate enough to synchronize and coordinate the parallel charging.

Sources:

1. Pop MSE, Lacho, Dimi Avram MSE. 2016. The New Simple and Practical Solar Component Guide, Kindle Edition. Digital Publishing Ltd

2. Morningstar Corporation. Morningstar's TrakStarTM MPPT Technology & Maximum Input Power - Morningstar Corporation [Internet]. [cited 2017 Dec 31]. Available from:
https://www.morningstarcorp.com/whitepapers/morningstars-trakstar-mppt-technology-maximum-input-power/

Inverters

The primary function of off-grid inverters is converting the output voltage of the battery bank to AC voltage. Not every off-grid solar system needs an inverter. An inverter is not needed if power is to be provided to DC loads only:

a) An inverterless off-grid photovoltaic system with a battery bank:

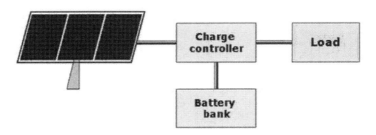

b) An inverterless off-grid photovoltaic system without a battery bank:

Important!

Grid-tied and off-grid photovoltaic systems use different kinds of inverters.

There are two types of off-grid inverters.

The first type is the off-grid inverter directly connected

to the solar array, thus providing power directly to the AC loads:

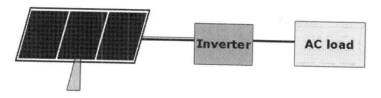

The second type is the battery backup inverter connected to the battery by a DC breaker:

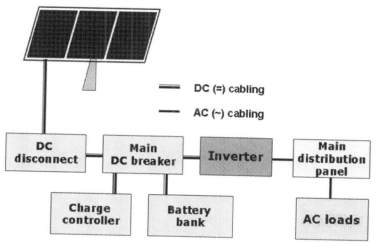

For efficient conversion of the DC power to AC power, the inverter's input voltage range must match the voltage range of the battery bank. The battery bank voltage reaches the lowest value when the batteries are discharged and the highest value when the batteries are fully charged.

Furthermore, an off-grid inverter usually comes equipped with a low-voltage sound alarm warning you that the battery voltage is about to drop below a critical point of discharge. Once this point is reached, the inverter starts shutting down to avoid any further discharge which is dangerous for your batteries.

There are three different types of standalone inverters currently available on the market, with regards to the produced type of voltage wave – sine-wave, quasi (modified) sine-wave and square-wave ones.

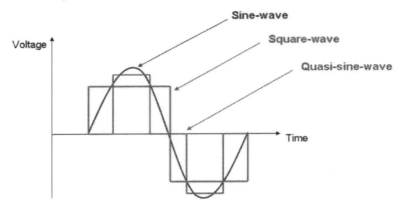

Although being the most expensive, sine-wave inverters are the best choice, since they are suitable for any applications and match best to the regulatory requirements.

Important!

Have in mind that some electronic equipment, such as mobile phones, microwave ovens, computers, vacuum cleaners, etc., might have problems operating with quasi sine-wave inverters. Furthermore, quasi sine-wave inverters might create additional noise for audio and TV equipment.

Inductive loads, such as fridges, pumps, drills, etc., must be powered by a pure sine-wave inverter. Furthermore, an inverter must be capable of providing a start-up current for such loads, which is usually 2 to 3 times higher than their nominal operating current.

Square-wave inverters are of worse quality than quasi sine-wave ones.

Important!

The most important features of off-grid inverters:

- Generation of a stable enough, sinusoidal AC voltage.

- Ability to provide enough power to all the connected electrical devices.

- Ability to withstand electrical surges created by loads containing motors.

- Low-energy consumption in the standby mode.

- Withstanding to tolerances in the battery voltage.

- Alerting when the battery capacity is low.

- Battery charging – converting the AC voltage generated by an external backup generator (optional) to DC voltage for charging the battery bank.

- Overcurrent protection.

With the idea of protection, off-grid inverters are designed to switch off automatically when the battery voltage falls below a certain level. Therefore, in an off-grid battery-based system, the battery charging level should be regularly checked.

Often an off-grid inverter and charge controller are combined into a single device. This leads to lower costs of building the solar panel system. Such a concept, however, has a notable drawback. Since a couple of devices are integrated into a single one, the system designer has less freedom, both upon sizing and selecting the components.

Part of the DC power generated by the solar array is

used for battery charging, and if the available capacity of the battery is sufficient, the other part of the DC power is turned into AC power to feed the household devices and other loads.

When the battery capacity reaches a specific minimum permissible level, the inverter could either alert the system operator to switch on an additional generator (operating on wind, diesel fuel, etc.) or itself automatically start such a generator to prevent overdischarging of the battery bank.

Off-grid inverters are produced in various power outputs, depending on the type and size of the solar panel systems. There are 100 W inverters for small standalone systems, and there are 5 kW inverters intended to power all the possible loads in residential solar power systems.

Another essential feature of battery-based inverters is that their DC input is available just for a limited number of DC voltages (12V, 24V and 48V), due to the reason that the inverter input often is also the battery output that comes in these DC voltages.

Specifications of off-grid inverters

- Rated input power – usually selected 20% lower than the solar array peak power, due to the losses in the solar modules.

- Rated output power – should be enough so that the inverter can handle all the loads working simultaneously.

- DC input voltage from the battery bank – the voltage values are standard, with the most typical ones 12V, 24V and 48V.

- Output voltage – usually 120 VAC or 230 VAC for most residential buildings.

- Output frequency – 50Hz in Europe, 60Hz in the USA.

- Surge capacity – allows the inverter to supply much more output power than its rated value within a short period, to provide a high starting current to motors (in refrigerators, water pumps, etc.).

System voltage

System voltage denotes the nominal voltage at which the solar array, batteries and charge controller operate. Often the system voltage is also the voltage at which some devices powered by the solar panel system operate. Here are some commonly valid general rules:

- Most small off-grid and mobile systems operate at 12V.

- Many charge controllers and off-grid inverters available on the market can operate at either 12V or 24V. Some of them are capable of sensing and adjusting the system voltage automatically.

- MPPT charge controllers are capable of accepting a range of voltages from the solar array (for example, between 20 and 100V) and deliver 12V or 24V to the battery.

- A system voltage of 24V is suitable for solar systems comprising long cable runs. Devices operating on 24V DC, however, are not widespread.

Other solar power system equipment

Cables are used to connect the individual components of an off-grid solar panel system. Cable is a wire put inside a conduit or pipe for protection.

A clear distinction should be made between the DC cabling and the AC cabling in a solar panel system.

The DC part of the cabling comprises the outdoor laid cables and the wiring between the modules (gathered in junction boxes, in case that the solar array comprises a couple of strings), between the strings (gathered in combiner boxes) and the connection to the inverter. To provide protection against ground-faults and short-circuits, the positive and negative poles of the cables should always be separated from each other. Furthermore, DC cables should be sheathed against unfavorable weather conditions, screened against lightning strikes, and mechanically protected.

AC cabling connects the inverter to the loads. AC cables do not need to be designed for outdoor conditions.

Important!

Higher system voltage results in lower currents, which means lower voltage drops in cables and hence, low cable losses. Higher currents require cables of a larger cross-section, which means a more expensive system.

Therefore, try to make the cable run between the solar array and the controller/batteries as short as possible.

The DC cabling, compared to the AC cabling, should have the following additional features:

- o Double insulation
- o UV-resistant and water-resistant
- o Designed for operation in wide temperature range (-40 to +120°C)
- o Designed for high voltage (more than 2 kV) operation
- o Easy to mount, light and flexible
- o Fire-resistant and low-toxic
- o Sized for low voltage drops.

- All the DC cables should be resistant to UV light and suitable for outdoor use.

- The DC cable losses (i.e., the voltage drop on the cables) should be kept down to a minimum, as follows:
  - o Less than 3% between the solar array and the battery.
  - o Less than 5% between the battery and the DC loads.

If the cross-section of the DC cables spanning between the solar panels and the battery is not large enough, the battery lifespan can be reduced significantly, especially at high ambient temperatures. The reason is that the maximum allowable voltage drop on the cables will be exceeded and the battery will not be able to get fully charged.

When more than two solar modules are needed, a DC breaker should be installed between the solar array and the charge controller, so that the system would be capable of disconnecting itself from the solar panels in case of repair works or during thunderstorms.

The most important electrical safety element in off-grid systems is the battery main fuse. It is intended to protect the battery from short-circuiting and must be mounted as close to the battery positive pole as possible. The battery main fuse should also be appropriately sized to protect all the components that can get affected by short-circuiting and overloading.

Short-circuiting between the battery poles is extremely dangerous. In case of short-circuiting, a very high current starts flowing (as a result of the very low internal resistance of the battery) which can result in a voltage arc – the cables can get melted, the battery can explode and dissipate sulfuric acid all around. Sulfuric acid is very hazardous to human health and can cause blindness, severe skin burn and lung damage. Therefore, make sure to provide suitably sized fuses or automatic breakers to each element connected to the battery.

Important!

1) The cables must be sized to conduct the maximum or short-circuit current of a circuit for at least 3 hours.

2) The cable size must be chosen taking into account the working temperature, i.e., the cable ampacity must be derated according to the operational temperature. The cable ampacity is rated at a given working temperature. The higher the working temperature than the rated one, the lower the cable ampacity as a result of the increased heat resistance of the cable. For example, a copper cable might conduct up to 50A at 60°C (140F), while only 41A at 75°C (167F). Therefore, if the

ampacity of the circuit required is 45A at the working temperature of 75°C (167F), you should select a cable type of a higher diameter.

3) The ampacity rating of the breaker and fuses should not exceed the cable ampacity rating but should always be similar to the pertaining short-circuit current. It's the breakers and fusses that are expected to protect the cable and the components, not the opposite!

4) The fuse size in amps must not exceed the cable ampacity.

Designing an off-grid solar panel system

Evaluating the solar resource

Even if your area and location have an excellent solar potential, your house might not be suitable for installing a photovoltaic system.

Your house is ready for installing a solar system if:

- You have already made it energy-efficient.
- The roof is unshaded, at least during the sunny hours (typically six) of the day.
- The roof has a Southern (or Northern, if you live in the Southern hemisphere) orientation.
- The roof is in excellent condition.

How to assess your location for the solar resource?

The spot where you intend to install the solar array should have:

a) Clear and unobstructed access to the sun throughout the day (between 9 a.m. and 3 p.m.) and throughout the year.

This means lack of any obstacles between the sun's rays and the solar array's surface – trees, chimneys, lamp-posts, neighbor buildings, etc.

It should be noted that a spot may be unshaded during one part of the day and shaded during part.

Furthermore, a site unshaded in summer might be shaded in winter, as the low position of the sun in winter casts longer shadows.

b) A preferably a South-facing (or North-facing, if you live in the Southern hemisphere) roof.

A full South (or full North, if you live in the Southern hemisphere) orientation is not mandatory. A somehow Southeast or Southwest (for the Southern hemisphere Northeast or Northwest, respectively) facing roof is also acceptable.

It has been proven that a deviation within 20-30 degrees of the full South (the full North, if you live to the South of the Equator) results in a degradation of solar array's performance of less than 10%, which is acceptable.

Important:

Pure Eastern or pure Western orientation is not recommended, since as a rule, solar modules should be exposed to direct sunlight for at least 6 hours a day.

You should mind that installing a solar panel system on a roof facing East or West might result in **20% degradation of system performance**, which is a severe compromise!

The roof can be either sloped or flat. A flat roof will allow you more easily to adjust the desired tilt of the solar array, but a sloped roof will be okay as well.

c) Enough space for placing the solar modules

The area you need for your solar system depends mainly on:

- How much energy it is designed to produce.

- The type of the solar modules you are going to install (monocrystalline, polycrystalline or thin-film).

- The size of the solar modules.

Important:

The less efficient modules you use, the larger the area you need for your solar array but also, the lower the costs.

Monocrystalline modules are the most efficient solar module type, while thin-film modules are the least efficient.

Why is surveying your site important?

Performing a site survey is the starting point for launching any solar panel system.

Important!

The electricity generated by a solar panel system depends on:

- Geographic location
- Orientation and tilt angle of the solar array
- Shading
- Installed photovoltaic power
- System efficiency.

Geographic location is crucial for the irradiation a solar panel system can obtain from the sun.

Orientation (azimuth) and tilt of the solar array are often determined by the orientation and slope of the roof where the array is mounted.

Shading is detrimental to any solar electric system, since it can dramatically bring down the system performance.

If you want your solar panel system to achieve the maximum performance, you should start with the first three of the above-listed issues.

How to perform a site survey by yourself

Performing a site survey is the starting point to launching any photovoltaic system.

Important!

When searching for a proper site to install the solar panels, you should consider the following:

- Orientation towards the sun
- Lack of any sources of shading (during the day and throughout the year)
- Minimization of the length of the DC cables spanning between the solar array and the inverter
- Protection from theft and vandalism
- Easy access for installation and maintenance of the solar array.

Below you can find described the five steps of a solar site survey you can perform by yourself.

Step 1: Assess any possible shading by nearby objects

The solar array should be provided with clear and unobstructed access to sunlight between 9 a.m. and 3 p.m. every day, throughout the year.

Important:

Shading is not recommended, at least between 9:00 a.m. and 3:00 p.m. Have in mind that even small shadows can severely affect the power output of the solar array.

To achieve the maximum of your shading analysis, perform the survey during a bright and sunny day, preferably in summer when trees have their full foliage mass.

During the site survey, you should be looking for the following obstacles:

- Buildings – you should be informed whether a new building is not being planned nearby, throwing shade to your site.

- Chimneys, power lines, poles, hedges, and neighboring roofs.

- Trees – if you are performing your site survey in winter, remember that in summer trees look different.

- Hills and other earth obstacles – mind that in winter the sun is much closer to the horizon than in summer.

Step 2: Determine the area available on your roof

Usually, the access space around the modules adds up to 20% to the required area for placing the solar panels.

Important:

Don't try to use every last square inch on your roof to install a solar array because:

- The array gets challenging to install.
- The array gets hard to maintain.
- Wind loading at the edge of the roof increases.
- It is possible to violate some regulations for provision of enough space to firefighters and other staff that might need access to the roof area.

Consider the dead spaces around the array. These are the spots that are either shaded or are to be provided between the modules.

Step 3: Determine the azimuth and tilt

Use a compass to check what direction your roof faces. Use a spirit level to measure the angle of the roof from the horizontal.

If your site is located in the Northern hemisphere, you should look towards the South, East, and West. If your location is in the Southern hemisphere, you should look towards North, East, and West. If you live near the equator, you should look towards East and West.

The ideal roof for mounting your solar array is a roof facing South if you live in the Northern hemisphere (and facing North if you live in the Southern hemisphere).

Things, however, are not as crucial as they appear. You can get 90-95% of a solar module's full power if it is located within 20 degrees around the sun's direction, which means 20 degrees to the East or to the West of the full South (or the full North, if you live in the Southern hemisphere).

Important:

Recommended limits for mounting the array:

Azimuth: maximum 30 degrees East or West from the South (or the North, if you live in the Southern hemisphere).

Tilt: ±15 degrees from the latitude of your location.

Step 4: Choose a mounting method of the solar array

There are the following types of mounting methods:

- Sloped roof mounting
- Flat roof/ground mounting
- Roof-integrated mounting
- Wall mounting.

Off-grid solar panel systems are rarely mounted on roofs. They are installed most often in rural areas, either on a ground-mounted construction or a pillar. Both flat roof mounting and ground mounting, however, give you the most freedom to optimize the position of the solar array according to the solar resource. There are four types of flat-surface mounted racks:

Fixed racks:

- Fixed at one orientation facing full South (or North, for the Southern hemisphere)
- Slope (tilt) equal to the site latitude.

Manually adjustable racks:

- Allow changing the tilt
- Usually, the tilt is changed at the start of every season
- Result in 12% power increase compared to fixed mount system.

Single-axis tracking racks:

- Follow the sun from East to West every day
- Require additional components and maintenance
- Increase the power output by 25% compared to the fixed mount system.

Dual-axis tracking racks:

- Continually orient the solar modules perpendicular to the brightest part of the sky.
- Require additional components and maintenance.
- Are more common than single-axis tracking racks.
- Increase the power output by more than 30% compared to the fixed mount system.

The other types of mounting give you less freedom, since you are not able to orientate the solar array to receive the maximum irradiation. The azimuth, however, is more important than the tilt, so evaluating the site from a solar point of view gives you an opportunity for compromise.

Important:

Regarding various mounting constructions, mind that not every mounting construction is suitable for every panel, while certain panels are designed for a specific mounting.

Step 5: Choose the azimuth and tilt of the solar array

Stationary mounted solar arrays should have an azimuth facing the average position of the sun.

Important:

- If you live in the Northern hemisphere (e.g., the USA, Canada or UK), solar modules should point as close to the full South as possible.

- If you live in the Southern hemisphere (e.g., Australia, New Zealand or South Africa), solar modules should point as close to the full North as possible.

If your roof, and the solar array respectively, are not oriented towards the full South (full North, if you live in Australia, New Zealand or South Africa), your solar panel system is about to lose 1.1% of its power output per every 5 degrees of deviation.

Example:

If your roof is oriented 30° from the full South (full North), the power output of your system will be reduced as follows:

Power loss, % = (30 degrees ÷ 5) * 1.1 =
= 6 x 1.1 = 6.6%,

which is acceptable.

You will experience the highest loss of power output if your roof faces the full East or the full West. Since the full East and the full West are 90 degrees away from the full South (full North), in such a case your system will lose:

Power loss, % = (90 degrees ÷ 5) * 1.1 =
= 18 * 1.1 = 19.8%

of its power output. 20% is a significant loss of power!

Important!

A solar array facing the full East or full West is assumed as <u>the worst</u> orientation, and must be avoided!

The tilt has less impact on the performance of a solar array than azimuth. A minimum tilt of 10° is recommended to ensure a surface self-cleaning by rainfalls.

You can get the optimal tilt for your site from the NASA solar database by using the following link:

https://eosweb.larc.nasa.gov/cgi-bin/sse/grid.cgi

https://asdc-arcgis.larc.nasa.gov/sse/

Or use this solar tilt free application for Android mobile phones:

https://play.google.com/store/apps/details?id=com.cl arkgarrett.solartilt&hl=en

Important!

Recommended tilts:

Tilt angle value	Performance target
= site latitude	For maximum performance throughout the whole year
= site latitude − 15°	For maximum performance in summer
= site latitude + 15°	For maximum performance in winter

Deviations from the above tilt values are acceptable in a wide range. Therefore, and also for aesthetic reasons, the tilt of the array can be made equal to the roof slope pitch, without any fear that the system will underperform significantly.

Source:

Pop MSE, Lacho,Dimi Avram MSE. 2015. The Ultimate Solar Power Design Guide: Less Theory More Practice, Kindle Edition. Digital Publishing Ltd.

Selecting the system components

Selecting the solar panels

The power output of a solar panel depends on:

- The number of solar cells in the panel
- The type of the solar cells
- The surface area of the cells
- The amount of solar radiation
- The tilt angle of the panel
- The temperature of the solar panel
- The voltage at which the load or the battery is drawing power from the panel.

The amount of radiation, the tilt angle, and the temperature of the solar panel all depend on where the solar panel is mounted. The voltage at which the load or the battery is drawing power from the panel depends on the battery, the charge controller, and the load. The idea is to select the components of the solar panel system so that to get the most power from the sun.

Solar modules should be selected to match the battery voltage. For most small and medium off-grid solar systems, the system voltage is typically selected either 12 VDC or 24 VDC. For large off-grid systems, it is often 48 VDC.

The operating voltage of the solar modules should be high enough to charge the batteries. A 12 V battery needs a charging voltage of 14.4 V. Solar modules should be capable of providing such voltage to the battery, after considering all the system losses – in cables, charge controller and diodes – and sometimes at high ambient temperatures. To ensure reliable charging of a 12V battery, a solar module of average Voc of about 20 VDC is needed.

The modules should contain enough solar cells connected in series to provide the necessary voltage. For 12V batteries, this means crystalline modules of 36 cells. For 24V batteries, modules of 72 solar cells (or 2 modules x 36 solar cells, connected in series) are needed.

The number of the solar panels needed for your solar system should be chosen based on the load analysis that is a must before starting to size the system. Information how to perform a load analysis is given in the "*System sizing Example 1: Sizing a solar system for a summer house, Step 3) Load analysis*" further in this book.

Here are some practical rules to be used upon the solar panel selection:

- If you are going to use crystalline solar panels, consider the number of cells in a panel. 36 cells in a panel are regarded as optimal. Panels comprising fewer than 34 cells are not recommended. For a system voltage of 24V and a battery bank of 24V respectively, 72-cell panels are also acceptable.

- Cost-effective 60-cell solar panels are not well suited to operate well with PWM controllers. Therefore, if you decide to use a PWM controller, you should choose 72-cell panels. You can only benefit from 60-cell panels if you use an MPPT controller.

- Check whether the solar panels are suitable for charging 12V batteries.

- To size a solar system, you need to know the main parameters of the solar panels – peak power, short-circuit current, open-circuit voltage, nominal working current. Be sure to have all these parameters stated. If you have a solar panel but you have lost its datasheet, you can find these parameters written on its label providing information also about the solar panel's type, certification, and manufacturer.

- When buying the panels, check whether they are suitable for off-grid systems. Certain solar panels are designed for grid-tied systems and cannot be used in off-grid systems.

- Check the guarantee offered with the panels. Have in mind that some crystalline modules come with 25 years of guarantee. Do not buy a panel offered with less than 5 years of guarantee!

- If you intend to buy secondhand solar panels, checking them with a multimeter is a must. For more details, please, refer to our book *'The Truth About Solar Panels: The Book That Solar Manufacturers, Vendors, Installers And DIY Scammers Don't Want You To Read.'*

- When selecting a solar panel, read carefully the technical parameters rather than trust any advertising-like statements. For example, don't let yourself be convinced by "This module is made for use in hot climates" but have a look at the I-V curves for temperature dependence instead.

- Last but not least – if in doubt, do not hesitate to ask an expert. Buying the wrong solar panels for your solar system is an expensive mistake that can be and must be avoided.

Source:

Pop MSE, Lacho, Dimi Avram MSE. (2015-10-26).The Truth About Solar Panels: The Book That Solar Manufacturers, Vendors, Installers And DIY Scammers Don't Want You To Read, Kindle Edition. Digital Publishing Ltd.

Selecting the batteries

Batteries often appear as the bottleneck of a solar power system, since they are bulky, costly, challenging to maintain and have a relatively short lifespan. Most off-grid solar electric systems are battery-based, so you should be well informed about the essential criteria for battery selection.

A poorly operating battery can not only reduce the performance of your solar system but can also damage your household devices and appliances. To choose the right battery means not only to find the battery matching technically to the rest of the system components but also to make the right decision to invest more money on a costlier battery that will be worth your investment over time.

Please, remember that not only batteries of a specific type but also batteries of a particular model might have, although slightly, different parameters. Therefore, any battery guide is just a leaflet on battery basics. Make sure you have all the specifications and guidelines available for the batteries you are buying. No book can replace the specifications and guidelines for battery installation, commissioning, maintenance, and replacement than the documentation provided by the manufacturer.

Do not trust plain declarations like "This is a solar battery" or "This is the best battery for your solar panel system". Instead, as is the case with solar panels, read the specifications and the technical parameters and characteristics rather than trust various advertising-like statements.

Important!

Consider using sealed batteries – Nickel- and Lithium Ion (Li-Ion)

These batteries are very different to lead-acid batteries, since they are sealed, portable, maintenance-free, and are used to power relatively small devices. Also, compared with lead-acid batteries, they are much smaller, lighter and have longer lifespan.

Nickel batteries (Metal Hydride and Ni-Cad ones) can be left fully discharged for extended periods. Li-Ion batteries, however, can be damaged by deep discharges. Nickel batteries can operate in a broader temperature range, while Li-Ion batteries perform well in colder temperatures and can get damaged by heat.

You should select the battery not just as one of the system components. Choosing the best battery for your solar system is vital, since the battery affects both the cost and the overall system performance. You should start with making a quick research about what is available at the market.

Only in case of no other option available, you could use an SLI (automotive) battery for a small off-grid system (10Wp-1kWp). However, we recommend you to use a truck SLI battery (100-120Ah), rather than a car SLI battery (50-60Ah).

Some more tips on battery selection:

- In case of a large system, it is worth spending more on a good battery.

- Before ordering a battery bank, you should find a suitable room for it.

145

- If you feel that maintaining a battery bank is too challenging for you, you should select maintenance-free batteries.

- It is practical to use batteries available on your local market. Ordering batteries from thousands of miles away can be both expensive and time-consuming.

- Batteries with lead-antimony plates are more tolerant to deep discharge than lead-calcium ones.

- If you think your solar system is likely to move around, choose a battery that is both easy to transport and resistant to vibration.

Selecting the charge controller

Do not hesitate to invest in a good charge controller. This will both increase the battery lifespan and will improve the system performance.

You should choose a charge controller with the right current and voltage ratings. The most important ratings of every charge controller are its input current (from the solar array), output current (to the loads) and nominal voltage.

You should also make sure that the charge controller you have decided to buy is the right one for your battery. If you are going to use a battery type other than lead-acid, i.e., of a VRLA or gel type, check whether that charge controller is suited for it.

When installing the charge controller, follow the user manual. Keep the manual as you might need it during the device operation.

Which kind of charge controller is better – MPPT or PWM?

If your system is below 1,500 Wp installed solar power and is more prone to deep discharging and overcharging, the battery lifespan appears as a primary concern. Therefore, you may opt for a cheaper PWM charge controller at the expense of reduced generation of electricity. Such losses in production can be compensated by increasing the number of solar modules installed.

If providing more power to the system loads and the battery bank is more important than extending the battery lifespan, you may opt for the more expensive MPPT charge controller.

PWM controllers are very suitable for small wattage solar electric systems.

MPPT controllers are always more efficient in 'squeezing' more solar power from the solar modules. MPPT controllers lead to a higher power output of 20-30% in winter and 10-20% in summer.

Important!

In hot climates, MPPT controllers tend to lose their primary advantage – the high efficiency.

In hot climates, the efficiency of an MPPT controller is only 10% better compared to the efficiency of a PWM controller. In cold climates, an MPPT controller might increase its efficiency up to 40% more compared to a PWM controller.

For small systems, however, the additional cost of an MPPT controller could be money better spent on a higher number of solar modules to offset efficiency losses, in combination with a PWM controller which is less expensive.

Another issue worth considering is the average warranty period – it is 5 years for charge controllers and 20-25 years for solar modules. Therefore, investing in larger solar arrays, along with buying a PWM charge controller, might be the better option to go for.

Selecting the inverter

In off-grid solar electric systems, an inverter can be designed to power either a single AC device or all the AC loads to be plugged into.

A common mistake in inverter selection is choosing an inverter performing poorly with solar systems due to either low efficiency or poor wave shape. Mind that inverters are not only used in solar power systems. For this reason, you should look for a 'solar inverter' and not just for an 'inverter.'

Do not fix on an inverter that does not have a datasheet with the full information you need. Often this is a sign that such an inverter is a device of low quality and performance.

Another typical mistake is selecting a grid-tied inverter instead of an off-grid inverter.

The inverter must be sized to handle the peak electricity demand.

The inverter must also match the system voltage (i.e., the voltage of the battery and the charge controller). Inverters for 12V or 24V system voltage are the most common, while 48V inverters are used in larger solar power systems.

To select an inverter for your off-grid system, you

need to perform load estimation (or load analysis). The load estimation is related to the loads you are going to use and how long you will use them. The load estimation will also summarize what and how many AC devices you are going to use and which of them will be operating at the same time. The inverter must be able to handle all the AC loads that are to operate simultaneously. Furthermore, the inverter must be able to handle the surge of these loads.

When buying an inverter, it is also important to consider adding new AC loads to the system. You should also consider whether and which of them are expected to work at the same time. Last but not least, you should buy an inverter that can be repaired. We also advise you to investigate the possible ease of support of the inverter you are planning to buy.

Selecting the devices and appliances

Incandescent and halogen lamps are not a good choice for any off-grid solar panel system. LED lights are preferable due to some advantages – extended life, high efficiency, durability, and smaller size.

It is proved that choosing efficient appliances can cut the cost of your solar electric system by half. Examples of efficient appliances include:

- LED or fluorescent lighting instead of incandescent bulbs

- Laptop computers instead of desktop computers

- Appliances that generally have high-energy consumption but are specifically designed for solar power systems – refrigerators, washing machines, and pumps.

When designing your off-grid system, you should

make a list of the devices and appliances that are going to operate simultaneously and then find out the rated voltage and power of each one. Furthermore, when buying a new device, you should always read carefully the voltage and power ratings.

Appliances typically operate either on DC (12V or 24V) or AC (120 or 230V).

Here are some reasons why you should prefer DC:

- DC-only solar photovoltaic systems are less complex and easier to operate and maintain.

- DC devices are more efficient to power directly from a battery than through an inverter. This is because, on one hand, the inverter's efficiency is practically never 100% (typically around 90%, as a result of the inverter losses), and on the other hand, due to the further degradation of the inverter's efficiency in case of using motor-containing devices (refrigerators, pumps) generating significant surges when started.

- In case of inverter failure, all the AC devices will stop operating as well.

- Often DC devices are more efficient than AC ones. This is valid especially when a device is specially designed to be plugged into an off-grid solar panel system.

Important!

When you need a system providing AC power only?

- You do not have any DC devices available. Normally in such cases, your system should comprise an inverter.

- The cabling and the circuits in the building where you are going to supply the solar-generated electricity are already designed for AC 120V or 230V. Therefore, you should design your solar system with an inverter, while complying with all the power efficiency rules and recommendations.

- You already have some efficient AC devices which you intend to use in the future – a Hi-Fi audio system or a TV set.

- You are bound to use longer cables for the DC part of your solar system. By using an AC inverter, you may opt in for a higher system voltage, i.e., go for 24V or 48V instead of 12V. Choosing a higher system voltage will decrease the DC current along the DC cabling and will ensure a longer cable run.

- You want to save money on cables – by using a higher system voltage, you will save on cabling due to the lower cable cross section required.

Important!

How to select the loads for your off-grid system?

- Avoid any old, inefficient devices (old TV sets or desktop PCs) that have high consumption and can drain much faster your battery bank.

- Unless you intend to build a very large system, do not use electric devices that contain heating elements (cookers, hair-driers, irons or toasters).

- For off-grid solar systems, DC refrigerators are

a better option than AC ones. Use compressor-based 12VDC fridges rather than absorption ones.

- Mind that for music systems, the stated rating often denotes the maximum consumption of the device and not the average one. For example, a Hi-Fi audio system may have a rating of 150W, while the real consumption might turn out to be as low as about 50W.

- Try to use appliances that are provided with rechargeable battery packs, such as electric drills, screwdrivers, etc.

- Beware of phantom loads. Mind that that all electrical devices in a standby mode do consume electrical power!

Off-grid solar panel system sizing

The electricity generated by a solar panel system depends on:

- Geographic location
- Azimuth and tilt of the solar array
- Shading
- Installed photovoltaic power
- System efficiency.

You already know how to perform a solar site survey and find the best location for your solar power system.

The installed photovoltaic power denotes the maximum power a solar panel can produce. This power, however, is inevitably decreased by the system losses.

The system losses include the solar module power tolerance, losses caused by dust and soiling, mismatch of the solar modules connected in a string, influence of the ambient temperature, as well as any losses in the cables and system components (inverter, battery, charge controller).

The system efficiency denotes the actual, real performance of a solar panel system, after considering all the system losses.

Important:

Off-grid solar system sizing has two key milestones:

1) Evaluating the battery capacity required

2) Assessing the installed solar power needed

1) Evaluating the battery capacity required to meet the daily energy target

$$Q = (Eday * DoA * BTM) \div (DoD * CE * Vsys)$$

Where:

- **Q** is the Minimum battery capacity required (in Ah)

- **Eday** is the Average daily energy target (in Wh) estimated by the load analysis of the available electrical devices used on a daily basis, considering their power ratings and periods of use.

- **DoA** is the Days of Autonomy (holdover) selected according to how important for you is to power the available appliances with no outages. The commonly assumed values are between 3 and 5 (days). If DoA<3, the battery will have a shorter life. If DoA>5, this will increase the battery costs; in such a case, an additional power source (for example, a diesel generator) is recommended.

- **BTM** is the Battery Temperature Multiplier. The efficiency of every battery decreases as temperature goes down, which in turn results in a reduction of the available battery capacity:

Ambient temperature	Battery Temperature Multiplier
80°F / 26.7°C	1.00
70°F / 21.2°C	1.04
60°F / 15.6°C	1.11
50°F / 10.0°C	1.19
40°F / 4.4°C	1.30
30°F / -1.1°C	1.40
20°F / -6.7°C	1.59

- **DoD** is the permissible depth of battery discharge. Any lead-acid battery must not be fully discharged. The DoD is given in the battery specifications. If the maximum depth of

154

discharge is 80%, then DoD=0.8. Although 80% is a commonly used value for lead-acid batteries, assuming DoD=0.5 (50%) will result in a longer battery lifespan.

- **CE (Cable Efficiency)** denotes the losses in the cables between the battery and the charge controller, and between the battery and the loads. If the total losses in these two cable branches are assumed 4%, then CE=0.96.

- **Vsys** is the System voltage (in V) that is 12V or 24V for small and medium off-grid solar system, and 48V for large systems.

2) Assessing the installed solar power needed

The daily energy output of a solar module can be calculated according to the formula:

E = Wp * PSH * SLF

Where:

- **E** is the solar module's energy output in Watt-hours (Wh)

- **PSH** is the daily Peak Sun Hours

- **SLF** is the System Loss Factor.

This formula can also be expressed otherwise. Instead of solar module's energy output **E**, we can use the daily consumption **Eday** to obtain how much installed 'peak' solar power **Wp** we need to meet that target consumption:

Wp = Eday ÷ (PSH * SLF)

Often, for the sake of simplicity, the following assumptions are made for **SLF**, depending on the off-grid system type:

- **SLF=0.7** for the simplest off-grid systems.

- **SLF=0.65** for battery-based off-grid systems without inverter.

- **SLF=0.6** for battery-based off-grid systems with inverter.

For detailed information on system losses, refer to our book "The Ultimate Solar Power Design Guide: Less Theory More Practice" or go to the section "System sizing Example 2: Sizing a mobile solar system, Step 3) Calculating the system losses".

PSH (Peak Sun Hours) can be estimated by using the NASA solar database (refer to **Appendix B**).

In the NASA dataset, you can find the average annual PSH and the worst month PSH value of your site. To size an off-grid solar system, you need the **worst month averaged** PSH, rather than the annual averaged PSH used for sizing grid-tied systems.

Here is a link to the NASA solar database starting page:

https://eosweb.larc.nasa.gov/cgi-bin/sse/grid.cgi?

Sizing an off-grid solar panel system comprises the following steps:

1) Solar resource evaluation

2) System concept design

3) Load analysis

4) Battery sizing

5) Solar array sizing

6) Charge controller sizing

7) Inverter sizing

8) (*optional*) AC generator sizing.

System sizing examples

Example 1: Sizing a solar system for a summer house

The summer house is located in the vicinity of Des Moines, Iowa and will be used by its owner during the period between April and October, seven days a week. Mostly AC devices are going to be used, except an energy-saving DC fridge.

The correct direction of the system sizing is from the load to the solar array. In other words, we should start with determining the system load, i.e., the total energy consumption, and then size the other system components.

The picture below depicts the losses introduced by the components of the solar power system.

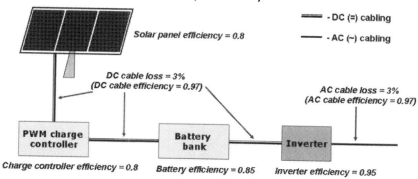

Step 1) Solar resource evaluation (refer to Appendix B)

Here is an extract from the NASA solar database concerning the area of Des Moines, Iowa:

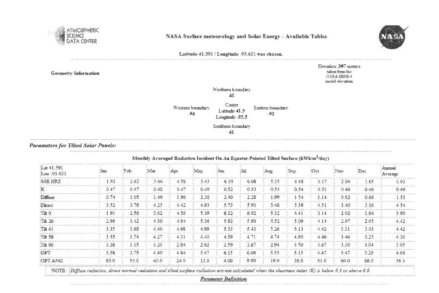

Geometry Information

Elevation: 397 meters
taken from the
NASA GEOS-4
model elevation

Northern boundary
42

Western boundary
-94

Center
Latitude 41.5
Longitude -93.5

Eastern boundary
-93

Southern boundary
41

Parameters for Tilted Solar Panels:

Monthly Averaged Radiation Incident On An Equator-Pointed Tilted Surface (kWh/m²/day)

Lat 41.591 Lon -93.621	Jan	Feb	Mar	Apr	May	Jun	Jul	Aug	Sep	Oct	Nov	Dec	Annual Average
SSE HRZ	1.92	2.62	3.66	4.58	5.43	6.10	6.08	5.35	4.48	3.17	2.04	1.65	3.92
K	0.47	0.47	0.48	0.47	0.49	0.52	0.53	0.53	0.54	0.51	0.46	0.46	0.49
Diffuse	0.74	1.05	1.49	1.96	2.30	2.40	2.28	1.99	1.54	1.14	0.82	0.66	1.53
Direct	3.52	3.76	4.25	4.42	4.93	5.73	5.95	5.48	5.38	4.51	3.40	3.16	4.54
Tilt 0	1.91	2.56	3.62	4.56	5.39	6.12	6.02	5.32	4.41	3.14	2.02	1.64	3.90
Tilt 26	2.96	3.42	4.30	4.84	5.36	5.83	5.90	5.52	5.09	4.14	2.97	2.65	4.42
Tilt 41	3.35	3.68	4.40	4.68	4.99	5.33	5.43	5.26	5.13	4.42	3.31	3.03	4.42
Tilt 56	3.55	3.74	4.27	4.31	4.40	4.59	4.71	4.74	4.90	4.46	3.46	3.25	4.20
Tilt 90	3.26	3.15	3.23	2.84	2.62	2.59	2.67	2.94	3.50	3.67	3.10	3.04	3.05
OPT	3.58	3.75	4.40	4.84	5.47	6.13	6.08	5.55	5.15	4.47	3.47	3.29	4.69
OPT ANG	63.0	53.0	40.0	24.0	12.0	4.00	9.00	19.0	36.0	51.0	60.0	66.0	36.3

NOTE: Diffuse radiation, direct normal radiation and tilted surface radiation are not calculated when the clearness index (K) is below 0.3 or above 0.8.

Parameter Definition

The system should be sized by considering the worst months for the solar resource, i.e., April and October. As we see here, PSH goes along with a particular tilt, which makes tilt selection easier. It is easy to find out that the most favorable combination of PSH for April and October is upon Tilt 41, where PSH is 4.68 and 4.42, respectively. Therefore, the system should be sized with **PSH=4.42**.

Step 2) System concept design

Since AC loads are to be used, an inverter is needed, along with a battery bank and a charge controller. The solar array will be placed in the yard, on a ground-mounted construction that is 30 meters (100 feet) away from the house. The solar array will be tilted at 41 degrees, as evaluated from above. Due to the relatively high cable run, a system voltage of 24V will be a good choice.

Step 3) Load analysis

The goal is to estimate the consumption of the individual electric loads. Remember, with an off-grid system above all you should search to use as many energy-saving loads as possible.

AC Loads	Power rating, W	Quantity	Total power, W	Hours of use/day	Avg. daily use, Wh
Indoor lights	11	3	33	4	132
Outdoor lights (CF)	40	2	80	0,3	24
TV	60	1	60	4	240
Laptop	60	1	60	7	420
Stereo Hi-Fi	12	1	12	5	60
Fan	40	1	40	3	120
Pump	500	1	500	0,5	250
Total AC power needed, W			785	Total daily AC use, Wh	1246
DC Loads					
Fridge	60	1	60	6	360
Total DC power needed, W			60	Total daily DC use, Wh	360

You can find the power rating on the label of each electrical device. By multiplying the power rating by the number of the devices of the same type used, you obtain the total power used by these devices. Next, you should estimate the average daily use or how long you use such devices every day. In case that a device is used not every day but once or twice a week instead, by dividing the total hours of weekly use by 7 (the days in a week), you get the average daily energy use of that device.

The total average daily energy use (for AC and DC devices) is calculated as follows:

$$Eday = Edayac \div (LFaccab * Inveff) + Edaydc$$

Where

- Eday is the total average daily energy use, Wh

- Edayac is the total average daily AC energy use, Wh. It is calculated by summing the average daily energy used by each AC device.

- Edaydc is the total average daily DC energy use, Wh. It is calculated by summing the average daily energy used by each DC device.

- LFaccable is the loss factor of cable losses in AC part of the system, typically 3%, LFaccable= 0.97.

- Inveff is the inverter efficiency, typically a value

159

between 0.85 and 0.95 (i.e., 85% and 95%). Here we assume a value of 0.9; for more details, refer to refer to our book *"The Ultimate Solar Power Design Guide: Less Theory More Practice"*.

For our example case,

Eday = Edayac ÷ (LFaccab * Inveff) + Edaydc
$$= 1{,}246 ÷ (0.97 * 0.9) + 360 = 1{,}787 \text{ Wh}$$

A general rule of a thumb concerning the Step 8 'Sizing the AC generator', is that you need an AC generator (e.g., a diesel one) when the daily energy use is above 2.5 kWh. Obviously, in this example, an additional AC power source is not needed.

Step 4) Battery sizing

We use the formula:

Q = (Eday * DoA * BTM) ÷ (DoD * LFdccab * * Vsys)

For this case,

- Eday=1,787 Wh.

- DoA – we assume 3 days.

- BTM – since the system will be used in summer, we assume BTM=1.

- DoD – we assume 0.5 (50%).

- LFdccab – efficiency of the DC cabling (the cabling between the battery and the loads), we assume 0.96.

- Vsys – system voltage, equal to 24V.

Thus

$$Q = (\text{Eday} * \text{DoA} * \text{BTM}) ÷ (\text{DoD} * \text{LFdccab} * \text{Vsys}) =$$
$$= (1{,}787 * 3 * 1) ÷ (0.5 * 0.96 * 24) = 465 \text{ Ah}$$

To implement this capacity, we have a couple of options available:

1) One 24V-battery of 500Ah

2) Two 24V-batteries of 250Ah each, and connect them in parallel:

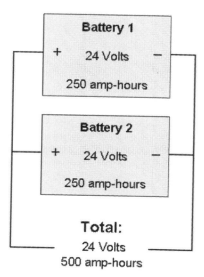

3) Two 12V-batteries of 500Ah each, and connect them in series:

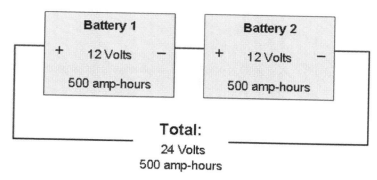

4) Four 12V-batteries of 250Ah each, and connect them in a mixed mode:

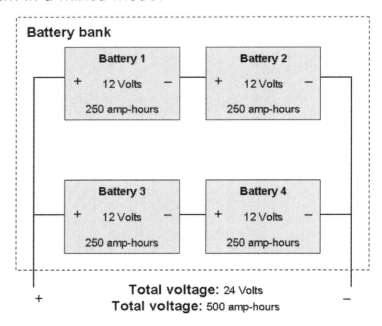

Which one of these options to choose depends on our budget and on what is available in the local market. It is worth, however, considering the following practical rules.

Important:

Connecting several low-voltage batteries in series is better than connecting several high-voltage batteries in parallel. Also, it is not recommended to connect more than 3 strings of batteries in parallel.

Considering this, for our example, we choose option 3):

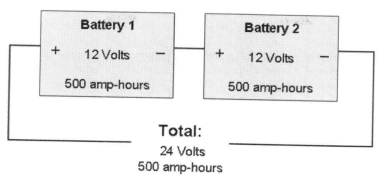

Step 5) Solar array sizing

If we use the above-shown formula

Wp = Eday ÷ (PSH * SLF),

And if Eday = 1,787 Wh, PSH = 4.42 and SLF = 0.6,

The peak solar power to be installed will be:

Wp = 1,787 ÷ (4.42 * 0.6) = 674 W

The number of solar modules needed can be obtained through dividing the Wp by the power rating of an individual solar module Wmodule:

N = Wp ÷ Wmodule,

and round the obtained value to the next integer number.

For example, if we select a solar module of 140Wp, we need:

674 ÷ 140 = 4.8,

That is, a total of 5 modules.

If we, however, decide to use a module of 310Wp rating, consisting of 72 cells, we will need fewer

163

modules of such type:

674 ÷ 310 = 2.17,

That is, a total of 3 modules.

A reasonable concept is to use 3 modules of 310Wp each for 24V nominal system voltage. For such modules, Voc=44V, Isc=9A, and Vmp=36V.

Since our system voltage is 24V, we cannot connect them in series unless we use a 'step-down'-featured charge controller. If we use a controller without such a feature, we have to connect all the 3 modules in parallel. Due to the low installed solar power of 930W, we may go for lower cost option when choosing the charge controller type, i.e., a PWM one. That is why it is recommended to go for 72-cell solar modules.

So we choose to connect these panels in parallel:

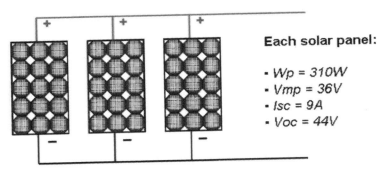

Each solar panel:

- Wp = 310W
- Vmp = 36V
- Isc = 9A
- Voc = 44V

Step 5) Charge controller sizing

For system sizing, the following parameters of the charge controller are the most important:

- Minimum input power **Pinmin**
- Minimum input voltage **Vinmin**
- Minimum input current **Iinmin**

- Minimum output current **Ioutmin**
- Maximum input voltage **Vocmax**

These parameters are calculated as follows:

Pinmin = Total solar modules x Wp of each module
Vinmin = Vmpcarray ÷ 1.25
Iinmin = Iscarray
Ioutmin = 1.25*Iscarray
Maximum input voltage Vocmax = 1.25*Voc

Vocarray and Iscarray are the open-circuit voltage and the short-circuit current of the solar array, respectively:

Vocarray = Voc x Number of modules connected in series

Iscarray = Isc x Number of modules connected in parallel

1.25 is the temperature derating coefficient.

For our case:

Pinmin = 3 * 310 = 930W
Vinmin = (36 * 1) ÷ 1.25 = 28.8V
Iinmin = 9 * 3 = 27A
Ioutmin = 1.25 * 27 = 33.75A
Vocmax =1.25 * 44 = 55V

As mentioned above, due to the low installed peak solar power of 930W, we can go for the lower cost option when choosing the charge controller type – i.e. the PWM one.

So we should look for a PWM charge controller of:

- Rated current ≥ 33.75A
- 24V nominal system voltage
- Maximum input voltage ≥ 55V
- Minimum input DC voltage ≤ 28.8V

165

Step 7) Inverter sizing

The inverter must be able to handle all the AC devices that are to be plugged into simultaneously (the AC total watts). The inverter must also be able to handle the expected surge ("in-rush" of current) produced by some large loads upon startup.

A common assumption for surge estimation is multiplying the total AC watts by 3. Most of the household devices, however, do not produce surges upon startup. Typical surging devices are refrigerators, washing machines, and pumps. When sizing the inverter, do not forget to compare the inverter's surge rating to the expected surge requirements of the system.

Other essential criteria when sizing the inverter are:

- Matching the inverter's input voltage with the nominal battery voltage.
- Selecting the desired AC output voltage (120 or 230 VAC).

For system sizing, the following parameters of the inverter are the most important:

- Inverter DC input voltage **Vinvin**
- Inverter AC output voltage **Vinvout**
- Minimum continuous output power **Pmincont**
- Minimum surge output power **Pminsurge**

Vinvin = System voltage

Vinout = the operating voltage of the AC devices

Pmincont = total power of the AC devices used

Pminsurge = (Total connected power of the surging AC devices * 3) + Total connected power of the non-surging AC devices

166

In this example, the only surging device is the pump consuming a total power of 500W. Therefore, the power of the remaining AC devices will be 785W – 500W = 285W.

For our case:

Vinvin = 24VDC
Vinvout = 120V
Pmincont = 785 W
Pminsurge = (500 * 3) + (785 – 500) = 1,785W

Given the above calculations, an inverter of 2,000W output power would be an excellent choice. It will be capable of meeting both the continuous and the surge power, and will provide us with the opportunity for possible adding of new AC loads (yet non-surging ones).

So, the solar panel system will comprise:

- Three 72-cells solar modules of 310 Wp and rated voltage 24V each;

- A battery bank of total capacity 500Ah consisting of two 12V batteries of 500Ah each, connected in series;

- 1 PWM charge controller of rated current of 35 A and designated for 24V nominal voltage;

- 1 inverter of 120VAC of 2,000W output power.

Here is a layout of the system we have designed:

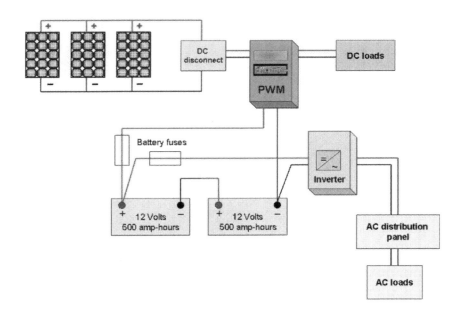

Source:

Pop MSE, Lacho,Dimi Avram MSE. 2015. The Ultimate Solar Power Design Guide: Less Theory More Practice, Kindle Edition. Digital Publishing Ltd.

Example 2: Sizing a mobile solar system

We are going to design a mobile solar power system designated for fulltime living in an RV vehicle in the vicinity of Iowa in summer and possibly in winter.

While sizing the solar system, we will pass through the following steps:

1. Designing the system concept

2. Load analysis (estimating the AC load consumption)

3. Calculating the system losses

4. Sizing the peak solar power Wp needed to install

5. Estimating the number of solar panels

6. Sizing the charge controller

7. Sizing the battery bank

8. Size the inverter

9. Sizing the overcurrent protecting devices: breakers and fuses

10. Sizing the cables

11. Sizing the transfer switch

12. Sizing the converter-charger.

Probably you have noticed that, compared to the previous example, new function blocks are introduced here – a transfer switch and a converter-charger.

To understand what their role in the solar power system is, let's get started with designing the system concept.

1) Designing the system concept

Our goal is to design a mobile solar system for fulltime living in an RV vehicle. The primary requirements for such a system are:

- To be small in size.

- To provide enough power for comfortable living in summer and possibly in winter.

- To be cost efficient – i.e., all electronic building blocks to be perfectly matched, and the system to be sized and optimized for maximum performance, thus squeezing the optimal power from the sun at any moment.

- To let charging the battery bank from an external source –a shore power or generator.

Here is a picture of the system:

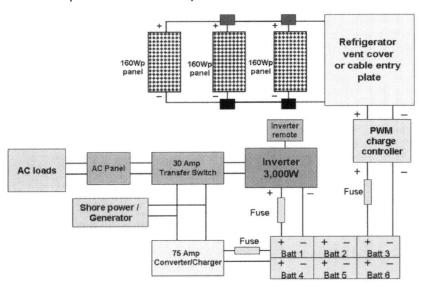

The system voltage will be 12V. The system is intended to power up various kinds of AC appliances and is not going to power any DC loads. Therefore, it needs an inverter to convert the solar-generated DC power into AC power. A battery bank is needed to store and convert the solar-generated electricity, and make it readily available at any time.

The mobile solar panel system will function as follows.

The solar panels convert the solar energy into DC electricity needed to charge the battery bank. The PWM charge controller protects the battery bank from overcharging by regulating the DC energy produced and sent to the battery bank. An inverter of 3,000 W output power converts the DC electricity stored in the battery bank into AC electricity of 120V or 230V voltage ready to power up the AC appliances. The transfer switch automatically detects and delivers AC power to the AC breaker panel. This AC power is generated either by the inverter or by an external generator, e.g., a shore power one. When the external/shore power generator is plugged into the system, the automatic switch automatically senses this external power and disconnects the solar generator. At that moment the switch starts delivering AC power to the appliances and converter/charger. The latter detects when the battery bank is not fully charged, and in such a case it is not, the converter/charger starts charging the battery bank. By automatically sensing and switching on and off between the AC power, the switch ensures that your AC appliances will always be powered up and the battery bank will always be charged via the converter/charger.

The converter/charger converts the external AC shore generator power to DC battery charging power and will charge the battery bank in case that the latter falls low in capacity. The converter/charger also regulates the battery charging, thus increasing the lifespan of the battery bank.

The inverter remote is a remote switch, which turns the inverter on and off. This switch is needed because the inverter is typically mounted in your RV vehicle so that it is usually hard to access.

A common mistake to avoid is connecting the converter/charger directly to the AC breaker panel. This will create a loop that will quickly discharge the battery bank, since the converter/charger will be charged by the battery bank via the inverter. To avoid this, always connect the converter/charger to the transfer switch.

2) Load analysis – estimating the AC load consumption

We assume that average daily consumption of 1,400 Wh is quite enough to ensure comfortable living in the RV vehicle. Such consumption allows us to use a variety of AC appliances every day.

Let's recall that the inverter size and the system size are determined by all AC loads running at the same time.

If we size the system with respect to the peak load, we will get a solar system that is both expensive and challenging to install. However, if we determine the system size with respect to the average load, yet considering the short-term running peak load, we would arrive at a system that is both less expensive and physically smaller.

For this reason, we assume average AC daily consumption of 1,400 Wh and a short-term peak consumption of 3,000W. Under 'short-term peak consumption' we understand maintaining the peak load up to 1h.

Therefore, we are going to size the system with

respect to the average daily load but with an inverter intended to sustain and meet the peak consumption.

3) Calculating the system losses

We are going to estimate the system losses to get the peak solar power (Wp) needed to install.

Let's list all the expected losses and derive the derating coefficients we need:

- The **solar panel losses** due to temperature and power tolerance are estimated 10%. Therefore, the derating coefficient Lsp=0.9.

- The **charge controller losses** depend on the charge controller efficiency. The charge controller efficiency is between 75% and 85% for a PWM controller and between 94% and 97% for an MPPT controller. Therefore, we assume the charge controller derating coefficient Lccpwm=0.8 for a PWM controller and Lccmppt=0.95 for an MPPT controller.

- We will use a PWM controller, since we are going to build a relatively low-power system.

- The **battery losses** are caused by the conversion of electrical energy into chemical and vice versa. The battery losses are about 80%. Therefore, the derating coefficient Lbat=0.85.

- The **cable losses** (reciprocally also known as **cable efficiency**) are caused by the resistance of the cable. We assume 2% losses in the DC cables and 2% losses in the AC cables. Therefore, the derating coefficients are Lcabledc=0.98 and Lcableac=0.98.

- The **inverter losses** are related to the inverter efficiency of converting the DC power into AC power. The average inverter efficiency is about

173

95%. Therefore, the derating coefficient Linv=0.95.

If we multiply the above derating coefficients, we will get the total system losses or the System Loss Factor (SLF):

SLF = Lsp*Lccpwm*Lbat*Lcabledc*Lcableac* *Linv = $0.9 * 0.8 * 0.85 * 0.98 * 0.98 * 0.95 = 0.56$
The System Loss Factor of the DC part of the system SLFdc is:

SLFdc= Lsp*Lccpwm*Lbat*Lcabledc =
$= 0.9 * 0.8 * 0.85 * 0.98 = 0.6$

The System Loss Factor of the AC part of the system SLFac is:

SLFac = Lcableac*Linv $= 0.98 * 0.95 = 0.93$

4) Sizing the installed solar power Wp

To determine the installed power Wp of the RV solar power system, we need to know the average daily Peak Sun Hours (PSH) for Iowa in summer and winter.

These are 5.3 in summer and 1.8 in winter (refer to **Appendix B** at the end of this book).

The nominal peak installed power Wp is:

Wp = Load ÷ (PSH * SLF) =
$= 1,400 ÷ (5.3 * 0.56) = 473$ Watt in summer and

Wp = Load ÷ (PSH * SLF) =
$= 1,400 ÷ (1.8 * 0.56) = 1,390$ Watt in winter.

For the sake of reducing the system cost, we are going to design the system with respect to the installed solar power needed in summer, i.e., 473W. The shortage of solar power in winter would be compensated by an external generator.

5) Estimating the number of solar panels

The number of solar panels N needed will be estimated by the formula:

N = Wp ÷ Wppanel

Where:

- Wp is the nominal installed peak power needed in Watts (W)

- Wppanel is the nameplate power of the specific solar panel.

We are going to use 160Wp solar panels. Therefore:

N= 473 ÷ 160 = 2.96

Since the calculated value N is always rounded up, we need 3 solar panels of 160Wp peak power each.

We decide to choose high-wattage 160Wp CTI-160 panels, since our RV vehicle has a limited mounting space on the roof.

The higher the solar panel wattage, the less space needed to install such a panel. What is more, using one high-wattage panel is usually more cost-efficient than alternatively deploying several low-wattage ones. Furthermore, such a high-wattage panel will provide us with the opportunity for any future system upgrades.

The specifications of the CTI-160 solar panel are:

SPECIFICATIONS

Rated power (Pm)	160W
Maximum power voltage (Vmp)	18.60V
Maximum power current (Imp)	8.60A
Open-circuit voltage (Voc)	22.50V

175

Short-circuit current (Isc)	9.29A
Power coefficient	−0.39% / °C
Voltage coefficient	−0.31% / °C
Current coefficient	0.045% / °C
Max. power tolerance	+/- 5W
Cell type	Monocrystalline
Module efficiency	16.4%
Series fuse rating	20A
Maximum system voltage	1000VDC (UL / IEC)

6) Sizing the charge controller

We are going to use a PWM controller, since our solar power system will be relatively small. You can use a PWM charge controller for solar systems of up to between 1,500 Wp and 2,000 Wp installed power.

Important!

The nominal voltage of a PWM charge controller must be equal to the nominal voltage of both the solar array and the battery bank. However, the nominal voltage of an MPPT charge controller voltage may be equal or higher than the nominal voltage of the solar panels and the battery bank.

In our case, the system nominal voltage will be 12V, and the voltage of the charge controller and battery bank will be 12V.

To keep the 12V nominal system voltage, we have to connect the three solar panels in parallel.

Upon sizing the charge controller, the essential solar panel parameters we should take into account are:

176

- Voc – the maximum open-circuit solar panel voltage at the lowest ambient temperature and the minimum open-circuit voltage at the highest ambient temperature.

- Isc – the solar panel short-circuit current at the highest ambient temperature.

Important!

The electrical specifications of the solar panel are given with respect to 25°C cell temperature. Therefore, the corresponding parameters should be derated in respect to the maximum and minimum ambient temperature.

Important!

When sizing a PWM charge controller:

- The charge controller must sustain the maximum solar panel current at the maximum ambient temperature.

- The maximum voltage of the solar panel (array) must be lower than the maximum input DC operating voltage. Otherwise, the charge controller might get damaged at the lowest ambient temperature.

- The DC voltage of the solar panel (array) must always be higher than the charge controller minimum DC operating voltage.

This rule will ensure that the PWM controller will always work and track the solar array at the highest ambient temperature.

Our solar array consists of 3 solar panels wired in

parallel. Therefore, the solar array voltage will be equal to the voltage of a standalone solar panel, while the solar array current will be a sum of the currents of the standalone panel.

Now, let's get started with sizing the PWM charge controller.

6a) Defining the charge controller current Icc

$$Icc \geq 1.25*Iscarray$$

Where

- Iscarray is the maximum short-circuit current of the solar array at the highest temperature.
- 1.25 is the temperature derating coefficient.

Therefore, for the charge controller current, in respect to our solar array consisting of 3 panels, we get:

$$Icc = 1.25*Iscarray = 1.25*3*Isc =$$
$$= 1.25 * 3 * 9.29 = 34.9A$$

Where Isc is the short-circuit current of the standalone solar panel.

Therefore, we can use a Morning Star PS 30 Pro PWM charge controller with the following characteristics:

- Rated Solar Current: 30A
- Rated Load Current: 30A
- System Voltage: 12V
- Min. operating voltage: 8V
- Maximum solar voltage for system voltage of 12V: 30V
- Available for positive or negative ground.

Mind that MorningStar charge controller current ratings are derated in advance. Therefore, we go for the 30A PWM controller.

6b) Checking the DC input operating window of the PWM controller with regard to the maximum and minimum operating voltage of the solar array at the highest and the lowest ambient temperature

A solar panel reaches its maximum voltage at the lowest ambient temperature. It is determined by Vocmax – the maximum open-circuit voltage of the panel. In our case, since we have 3 panels connected in parallel, the solar array's maximum voltage is equal to the standalone solar panel voltage.

Therefore, we can calculate the solar array open-circuit maximum voltage Vocarraymax as:

Vocarraymax = 1.25*Vocarray =
= 1.25*22.50V = 28V

Where

- Vocarray is array open-circuit voltage at 25°C.
- 1.25 is the temperature derating coefficient accounting for the lowest expected ambient temperature.

6c) Checking whether the minimum voltage of the solar array is always higher than the minimum DC input voltage of the PWM controller

The solar panel minimum voltage is the open-circuit voltage at the highest operating temperature Vocmin:

Vocmin = Vmpparray ÷ 1.25 = 18.6 ÷ 1.25 =
= 14.88V

Where

- Vmpp is the solar array maximum power tracking voltage, which in our case is equal to the solar panel standalone voltage Vmpp.

179

- 1.25 is the temperature derating coefficient accounting for the highest ambient temperature.

Again we are happy with our choice because 14.88V is higher than the PWM minimum operating voltage of 8V.

Important!

The temperature derating coefficient of 1.25 used here accounts for extreme high and extreme low ambient temperatures, i.e., the worst-case scenarios. For more accurate yet complicated determining of temperature coefficient at a given ambient temperature, please, refer to **Appendix D**.

If you are going to use an MPPT charge controller, please, refer to **Appendix E** on the guidelines for sizing an MPPT charge controller.

7) Sizing the battery bank

Before we start sizing the battery bank, let's remember some essential design guidelines:

- A string comprised of many low-voltage high-capacity batteries connected in series is a better option for getting the capacity required than a string composed of one or two high-voltage batteries connected in parallel.

- You should not connect more than 2 or 3 strings in parallel. Otherwise, the difference in the string voltage or a damaged battery in any of the strings can reduce the lifespan of the other batteries. Furthermore, it is difficult to maintain a uniform charging and discharging process per string, due to the inevitable small difference between the voltages of the individual strings.

- The battery bank capacity depends on the ambient temperature – the higher the temperature, the higher the capacity, but at the expense of a shorter battery lifespan.
- The higher the depth of discharge, the lower the cost of the battery bank, but at the expense of a shorter battery lifespan and fewer recharging cycles available.
- All the batteries forming a battery bank should be of same type, model and production series.
- The battery bank should be sized so as to allow charging to 100% at least once a week during the most unfavorable solar power conditions (days of poor sunlight).

The battery bank capacity is:

Q = (Eday * DoA * BTM) ÷ (DoD * Vsys)

Where:

- **Q** is the Minimum battery capacity required (in Ah)
- **Eday** is the Average daily energy target (in Wh), calculated by the load analysis assessing the electrical appliances used daily, considering their power ratings and periods of use.
- **DoA** is the Days of Autonomy (holdover)
- **BTM** is the Battery Temperature Multiplier
- **DoD** is the permissible depth of battery discharge
- **Vsys** is the System voltage (in V).

So for summer, DoD = 50%, BTM=1, and DoA=3, we have:

Q = (1,400 * 3 * 1) ÷ (0.5 * 12) = 700Ah

Therefore, we need a 12V battery of 700Ah. Let's consider using 6V batteries.

7a) Calculating the number of batteries in a string and the number of strings connected in parallel

The below equations will help us have the answer available on every occasion:

Battery string capacity [Ah] =
= Battery bank capacity ÷ Max. number of strings

Where 'Battery string capacity' is the capacity of a standalone battery connected in series to form a string.

So, in case of two strings, we will have:

Battery string capacity [Ah] = 700 ÷ 2 = 350Ah

Each battery should have a capacity of 350Ah, so we have to connect 2 battery strings in parallel. How many batteries, however, should we connect in series to form a string? To find the answer, we use the following formula:

Bs = System voltage ÷ Battery voltage

Where:

- Bs is the number of batteries connected in series to form a string.

- System voltage is the solar power system nominal voltage.

- Battery voltage is the voltage of a standalone battery.

So, for the number of batteries connected in series, we get

Bs = 12 ÷ 6 = 2

Therefore, we have two strings consisting of 6V batteries of 350Ah capacity each. Each string is composed of two 6V batteries connected in series.

The diagram below shows how a battery bank consisting of parallel strings is optimally connected to a charge controller.

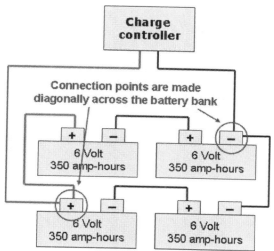

Mind that the optimal diagonal connection of the charge controller to the battery bank is essential, as it ensures more evenly distributed charging across the bank.

8) Sizing the inverter

As mentioned above, the following parameters of the inverter are essential for the system sizing:

- Inverter DC input voltage **Vinvin**
- Inverter AC output voltage **Vinvout**
- Minimum continuous output power **Pmincont**
- Minimum surge output power **Pminsurge**
- The generated output power should be a pure sine wave.

183

Vinvin should be equal to the system voltage and should fall within the recommended minimum and maximum battery bank voltage maintained during the conditions of fully discharge and charge.

Vinout should be equal to the operating voltage of the AC devices.

Pmincont is the total power of the AC devices used.

Pminsurge = Total connected power of the surging AC devices + Total connected power of the non-surging AC devices

So, for our case, we are looking for an inverter with the following characteristics:

- Inverter DC input voltage **Vinvin=12V**

- Inverter AC output voltage **Vinvout=120V**

- Minimum continuous output power **Pmincont>1,400W**

- Minimum surge output power **Pminsurge>3,000W sustained for up to 1h**.

The inverter surge power is the power to be sustained by the inverter for several seconds. Since we would like this power to be available for up to 1h, we should look for an inverter of 3,000W rated continuous power.

Here are some examples of inverters matching the above estimates:

1. GP-ISW2000

- Continuous Output Power 3,000 W (VA)

- Maximum Surge Rating 6,000 W (VA)

- Output Waveform: Pure Sine Wave

- Input Voltage 12 V: 10.5-16.5 VDC

2. Samlex SA-3000K-112

- Continuous Output Power 3,000 W (VA)
- Maximum Surge Rating 6,000 W (VA)
- Output Waveform: Pure Sine Wave
- Input Voltage 12 V

3. VertaMax 3,000 Watt 12V Pure Sine Wave Power Inverter

- Continuous Output Power 3,000 W (VA)
- Maximum Surge Rating 6,000 W (VA)
- Output Waveform: Pure Sine Wave
- Input Voltage 11-15VDC

9) Sizing the overcurrent protection devices – breakers and fuses

You should always select a fuse of ampacity rating either equal to the calculated value or of the next higher rating available.

We can use a fuse for each solar panel and a combiner with a breaker to combine all the 3 solar panels in parallel.

Also, we have to install fuses:

- between the charge controller and the battery,
- between the battery bank and the inverter, and
- between the converter/charger and the battery bank.

We are going to use an AC panel with a breaker between the inverter, the transfer switch on the one hand and the AC loads on the other hand.

Now, our task is to size the fuses to protect each standalone solar panel.

185

The solar panel fuses are sized for at least 3 hours of continuous operation by using an oversizing coefficient of 1.25, according to the NEC code, and an additional coefficient of 1.25, taking into account the irradiance variation as explained below.

9a) Sizing the fusses and breakers between the solar panels and the charge controller:

The DC current of the solar panel fuse is:

Ifuse = 1.56*Isc = 1.56*9.29 = 14.5 A

The fuse DC voltage is:

Vdcfuse = 1.25*Vocmax = 1.25*1.25*Voc = = 1.25*1.25*22.05 = 34.4V

We will use a combiner box to connect our 3 solar panels in parallel. The breaker included in this box should have the ampacity rating as follows:

Ibreaker = 1.56.*Iscarray = 1.56*3*9.29 = 43.5A

Vdcbreaker = Vocarraymax = 1.25*Voc = = 1.25*22.05=27.6V

9b) Sizing the fuse between the charge controller and the battery

The fuse ampacity rating is:

Ifuse = 1.25* Iccrated current = 1.25*30 = 38A,

where **Iccrated current** is the rated current of the charge controller.

The fuse DC voltage is:

Vfuse = Vbatmax = 15V

9c) Sizing the fuse between the battery bank and the inverter

The fuse ampacity sized in regard to continuous

power is:

Ifuse = 1.25*(Pacinv ÷ Vbatlow) =
= 1.25*(3,000÷11) = 341A,

where **Vbatlow** is the lowest battery voltage.

The fuse DC voltage will be:

Vfuse = Vbatmax = 15V,

where **Vbatmax** is the maximum battery voltage.

However, the surge power of the inverter according to the specification is 6,000W and is to be maintained for 3 seconds. If we connect a load requiring such power, the battery bank will be able to provide this wattage.

So, the fuse ampacity sized with regards to the inverter's 6,000W peak power will be:

6,000W ÷ 15V = 400A

In this case, you should use at least 4/0 cable gauge.

9d) Sizing the AC breaker of the AC panel after the inverter:

The breaker ampacity rating is:

Ibreak = 1.25*(Pinvac÷Vinvac) =
= 1.25*(3,000÷120) = 31A,

where **Pinvac** is the inverter continuous AC output power and **Vinvac** is the inverter AC output voltage.

The AC voltage rating of the breaker is 120V. On the other hand, we are going to use an automatic 30A-rated switch.

10) Sizing the cables

The minimum cable ampacity is sized similarly to the fuses, by using the same formulas per cable needed for the different sections described above, i.e., the cables connecting the solar panels, the cables needed for the section after combining the solar panels, etc. However, the corresponding cables should be sized to conduct the same or higher currents than the ratings of the fusses and the respective solar devices. Also, this current must be derated with respect to the cable operating temperature.

As already mentioned, the higher the working temperature of the cable in respect to the rated one, the lower the current it can conduct. Therefore, you might need a cable of a higher cross section (diameter) for operation at higher temperatures.

To size the cables, please use our calculator at: http://solarpanelsvenue.com/free-solar-cable-size-calculator/

11) Sizing the transfer switch

The transfer switch should be able to hold the continuous current for more than 3 hours.

Since the current in question is generated by the inverter, we can estimate it as:

Iats = Inverter power ÷ Output voltage =
= 3,000 ÷ 120 = 25A

Where **Inverter power** is the inverter continuous power and **Output voltage** is the inverter output voltage.

We may consider these 3,000W (which can be used for up to 1 h) as continuous power. So, in our case, we should use a transfer switch rated above 25A. We are going to use a 30A-rated one.

188

12) Sizing the converter/charger

The converter charger should deliver the optimal current for charging the battery bank when the system is connected to the shore power, and when the battery bank capacity falls sufficiently low. This charging current should be about 1/10 of the battery capacity.

In our case, the battery bank capacity is 700Ah, so the optimal charging current is about 70A at 12V DC.

The input of the converter/charger should be capable of operating with 120V AC shore power, while the output DC voltage should be high enough to charge the battery bank. Such a converter/charger is recommended to support a charging mode ensuring as extended battery lifespan as possible.

Sources:

1. Carmanah. CTI 160 Solar Module [Internet]. [cited 2017 Dec 31]. Available from: gpelectric.com/files/gpelectric/.../MOBI_SPC_CTI-160_vA.pdf

2. Morningstar Corporation. Morningstar-ProStar-Datasheet [Internet]. Available from: https://www.morningstarcorp.com/products/prostar/

3. Go Power! INDUSTRIAL PURE SINE WAVE GP-ISW2000 / GP-ISW3000 [Internet]. 2016. Available from: http://gpelectric.com/files/gpelectric/documents/PDF/SPEC_GP-ISW2000-3000.pdf

4. Samlexamerica. 3000W 12VDC-120VAC Pure Sine Wave Power Inverter | SA-3000K-112 [Internet]. [cited 2017 Dec 31]. Available from: http://www.samlexamerica.com/products/ProductDetail.aspx?pid=129

5. VertaMax 3000 Watt 12V Pure Sine Wave Power

Inverter DC to AC Car, RV, Solar, Off-Grid Applications [Internet]. [cited 2017 Dec 31]. Available from: https://www.windynation.com/Inverters/Windy-Nation/VertaMax-3000-Watt-12V-Pure-Sine-Wave-Power-Inverter-DC-to-AC-Car-RV-Solar-Off-Grid-Applications/-/1441?p=YzE9NDg=

6. Morningstar Corporation. Morningstar Solar Controller System Sizing with 60 Cell Modules - Google Search [Internet]. [cited 2017 Dec 31]. Available from: http://support.morningstarcorp.com/wp-content/uploads/2014/07/tech-tip-60-cell-PV-module-sizing.pdf

Installation of off-grid solar panel systems

Solar array

If you have solar panels installed on the roof, a good idea would be to use an inside, rather than an outside wall, to run the cable. If you have solar panels installed on a pole or on the ground, you should use underground cables.

- Always connect the modules to the batteries and the charge controller last in the sequence.

- Use correctly sized cables. To size the cables fast and easy, please, refer to our book *"The Ultimate Solar Power Design Guide: Less Theory More Practice"*.

- If earthing is required, you should provide a separate cable to each solar panel of the array, connecting it to a central earthing terminal.

- To avoid corrosion, make sure that the junction boxes of the modules are well sealed and carefully wired.

- The most common modules available on the market are for 12V, 36-cell crystalline ones. If your system voltage is 12V, the modules are likely to be wired in parallel. If you select a system voltage of 24V, then you have to wire two such modules in series to two 12V-batteries connected in series.

Batteries and charge controller

Be sure to install the main fuse close to the positive battery terminal. Leave one terminal disconnected from the battery until performing the final connection sequence. If your system comprises three or more batteries, a separate room for battery housing is recommended. In case of several batteries, make sure you connect them in the right way:

- In 12V systems, 12V batteries should be connected in parallel.

- In a 24V system, if you have two 12V batteries, you should connect them in series. If you have four 12V batteries, connect two ones in series and two ones in parallel.

During lead-acid battery charging, hydrogen is released. A gaseous mixture of hydrogen and oxygen from the air is highly explosive. Therefore, the battery room should be well ventilated. Furthermore, install the battery separately from the remaining equipment – inverters, controllers, circuit breakers. Such equipment can produce sparks leading to an explosion. On the other hand, the evaporations from the battery are highly corrosive and can damage the other electrical components.

Never disconnect the battery from the charge controller while the solar array is still connected to the controller. The latter might get damaged by the high open-circuit voltage of the array. Always disconnect the array from the controller before disconnecting the battery. During installation, always connect first the battery to the controller, and then the controller to the solar array. The fuses must never be switched in a live circuit – such action might result in igniting dangerous electric arcs.

Inverter

Inverters are rarely protected from a reversed polarity. If an inverter is connected to the battery by a reversed polarity, the inverter can get severely damaged. Usually, such an event is not covered by the inverter's warranty.

Regardless of who is going to install the inverter, the following rules are to be kept to ensure maximum safety:

- The inverter should be mounted high enough above the ground to eliminate access to it by children and animals.

- The inverter should be installed in a dry and well-ventilated place, with enough space available for heat dissipation.

- The inverter, along with its cabling, must be kept away from any flammable materials – gas, oils, solvents and other volatile substances that can easily be ignited by an accidental spark produced by the inverter.

Mounting the inverter in a small room, with not enough ventilation, near flammable materials can result in fire and even explosion!

Sources:

1. Antony, Falk, Christian Durschner, Karl-Heinz Remmers. 2007. Photovoltaics for Professionals: Solar Electric Systems Marketing, Design and Installation, Routledge.

2. Pop MSE, Lacho,Dimi Avram MSE. 2015. The Ultimate Solar Power Design Guide: Less Theory More Practice, Kindle Edition. Digital Publishing Ltd.

Maintenance of off-grid solar panel systems

Solar array

In most cases, no special cleaning of the photovoltaic array is needed. Usually, the rain is capable of washing the panels except in too dry and dusty regions. Small amounts of dust for a short period will not have any significant impact on the performance. Nevertheless, the solar modules should be cleaned from continuous soiling, such as one accumulated along the edges of the module frames or bird droppings. No solvents are needed; water with some soap is enough.

It should also be noted that, except for the most remote Northern or Southern regions, accumulation of snow on the solar panels when left unclean even for a couple of days does not affect the system performance significantly.

Batteries

Once you have a battery-based solar panel system, you should consider the following general recommendations:

- Never charge a battery without a charge controller.
- Never expose a battery to direct sunlight. Furthermore, the temperature of the battery should not exceed 50ºC (122F).
- Never fully discharge a lead-acid battery. Remember to design the system for maximum 50% daily discharge. Thus, you will provide a longer lifespan to your battery.

- Consider the deadline by which a battery needs to be replaced. Do not wait till the last moment; instead plan all the activities related to the battery replacement (checking for market availability, selecting a particular model, contacting the vendor, placing an order, transportation, etc.)

The charge current must never exceed 10% of the battery capacity. When batteries are charged at a current higher than 10% of the capacity, the electrolyte quickly 'evaporates' and the battery cells will get damaged. For example, the charge current of a 100Ah battery must not be higher than 10A. Lower charge currents (4-5% of the battery capacity) are better for battery charging.

If a battery is left uncharged for long, it will lose a part of its energy by self-discharge. Self-discharge depends on temperature, battery type, battery age, and storage conditions. Lead-antimony batteries have a higher self-discharge rate than lead-calcium batteries. High ambient temperatures increase the rate of self-discharge. To avoid self-discharge, store batteries on non-metallic surfaces, off the floor, and keep both the top surface and the terminals of the battery clean.

A typical cycle of maintenance activities, depending on the battery type, performed every 6 months, will include:

- Checking whether the connections are stable and not corroded – grease is an excellent protection.

- Checking the level of the electrolyte (for lead-acid batteries) and refilling with deionized/ distilled water, if necessary.

195

- Cleaning the battery surfaces from dirt, dust or moisture.

- Measuring the voltage of the cells/batteries – when fully charged and upon any loads and power sources disconnected for at least half an hour.

Thank you very much for reading this book!

If you like this book, we would be more than happy if you would leave a review. Your review does matter! It helps future readers and helps us improve the content of our book. Thank you in advance for this gesture of goodwill.

For updates about new releases, as well as exclusive promotions, visit the authors' website http://solarpanelsvenue.com where you can find many free solar calculators and other useful solar information.

Sign up for the VIP mailing list at:
http://solarpanelsvenue.com/authors.php

We need your valuable feedback. If, by any chance, you are disappointed with our book, please let us know directly via our personal email:

author@solarpanelsvenue.com.

We will work on your feedback and fix our mistakes in our updated version.

Also by the authors

The Ultimate Solar Power Design Guide: Less Theory More Practice **(The Missing Guide For Proven Simple Fast Sizing Of Solar Electricity Systems For Your Home, Vehicle, Boat or Business)** [Paperback and all types of eReaders editions – Kindle, Kobo, Nook, Apple, etc.]

ISBN-10: 6197258048
ISBN-13: 978-6197258042

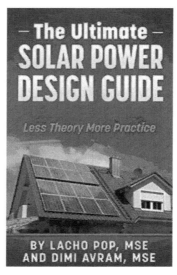

The Ultimate Solar Power Design Guide is a straightforward guide on solar power system sizing. It is written by experts for beginners and professionals alike.

Proper sizing of a solar panel system is crucial. The final goal of sizing a photovoltaic system is to come up with a cost-effective, efficient, and reliable photovoltaic system for your home, RV vehicle, boat or business – a solar panel system that squeezes out the maximum possible power for every cent invested.

The main drawback to the majority of solar books is that they provide too much general information about solar panels and solar components and, if you are lucky enough, just one or two fundamental sizing formulas.

The mission of this book is to fill this gap by offering a simple, practical, fast, step-by-step approach to sizing a solar panel system of any scale, whether simple or complex, intended for your home, business, RV vehicle or boat.

The book is written by experts holding master's degrees in Electronics, and it is targeted for those who cannot get started or are utterly confused.

It contains lots of fast and straightforward universal sizing methods applicable to all cases, accompanied by proper sizing examples. Thanks to this approach, you will be capable of sizing any solar power system or tailor the sizing methods according to your needs.

The New Simple And Practical Solar Component Guide [Paperback and all types of eReaders editions – Kindle, Kobo, Nook, Apple, etc.]

ISBN-10: 6197258099
ISBN-13: 978-6197258097

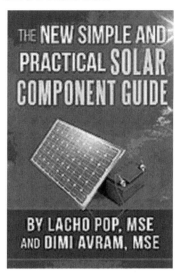

Have you ever wanted to save money on electricity and become energy-independent?

Do you want to protect your family from regular power outages and negligence of local utility?

The book "The New Simple and Practical Solar Component Guide" helps you accomplish this by understanding the essentials of building blocks of photovoltaics and harnessing solar power.

Written by electronic engineers, this easy-to-read-and-follow solar component guide demystifies all of the components of a solar power system in a way that anyone lacking technical background can understand.

The book is useful for a broad audience: technically and non-technically inclined people, beginners and advanced in solar power, and professionals of engineering background.

Based on hundreds of hours of research and experience, the book contains practical solar power information you cannot find and cannot apply by merely searching the web.

The Truth About Solar Panels: The Book That Solar Manufacturers, Vendors, Installers And DIY Scammers Don't Want You To Read
[Paperback and all types of eReaders editions – Kindle, Kobo, Nook, Apple, etc.]

ISBN-10: 6197258013
ISBN-13: 978-6197258011

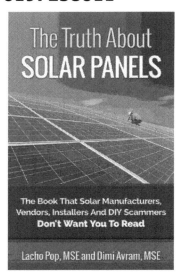

The book is about solar photovoltaic panels – the central building unit of solar systems. By reading this book, you get the complete know-how to buy, build, compare, evaluate, mix and assemble different types of solar panels.

It describes the basic solar panel types and reviews in details their main features and parameters. The book makes a unique presentation of cheap and second-hand solar panels by describing their pros and cons, where to find them, how to assess them and how to use them. Particular attention is paid to DIY solar panels by providing a proven methodology for evaluating a small solar electric system based on DIY panels. All the information is presented in an easy-to-read manner, with lots of examples, recommendations and tips, and proper summaries where needed.

The book is targeted to various groups of readers – homeowners, do-it-yourself solar enthusiasts, lovers of recreational vehicles, campers, boats and other outdoor activities, survivalists, potential investors in solar power, business owners interested in solar power, students, teachers, people interested in becoming solar installers, people working in the sales and marketing area of solar power and green energy industry and many others keen in solar power and renewable energy.

Top 40 Costly Mistakes Solar Newbies Make: Your Smart Guide to Solar Powered Home and Business [Kindle and Paperback Edition], Kindle ASIN: B01GGB7QP8, Paperback

ISBN-13: 978-6197258073
ISBN-10: 6197258072
ISBN-13: 978-6197258073

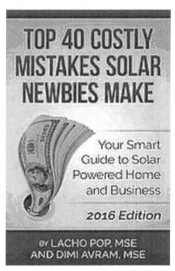

The book "Top 40 Costly Mistakes Solar Newbies Make" is a simple and practical guide that could save you a lot of money, headaches and time during the planning, buying, implementation, and operation phase of your solar power system.

Whether you have decided to buy a solar electricity system or assemble it yourself, you need a simple and easy-to-follow step by step guide.

There are thousands of books, articles, leaflets, forums and many other resources available online, written by unqualified authors, telling you what to do.

They could be misleading you!

By reading this book, you get trustworthy information

in the form of practical tips and hints given by engineers with long-term experience in electrical and electronic engineering.

Here are the most common groups of mistakes commonly committed by people who have decided to go solar:

- General mistakes and misconceptions
- Mistakes during location assessment
- Mistakes with solar panels
- Mistakes in solar system sizing
- Mistakes in assembling the system components
- Mistakes in buying a solar system
- Mistakes in solar power system maintenance.

Although solar electrics are everywhere around us and there are millions of resources available on this topic, you don't need to be an expert to assess what is right for your specific case. Neither you have to refer to costly solar consultants to reveal you the solar secrets.

What you need is this simple and easy-to-read solar book. It will save you both money and trouble down the road. It is frankly written by experts for everyone who cannot get started.

Get started your solar journey today!

Appendix A: Voltage, current, power, energy

Voltage is the difference between the potential ('hidden') energy between the two ends of a conductor. It is measured in 'Volts' and is denoted by the symbol '**V**.'

Current is the flow of electrons in a circuit. It is measured in 'Amps' and is denoted by the symbol '**I**.' The cause for current to start flowing is voltage.

A **circuit** is a combination of electrical devices and cables providing a closed path for electrons, thus making current to flow.

Resistance is the opposition of a conductor to electric current. Resistance is denoted by the symbol '**R**' and is measured in Ohms (Ω).

Energy is the ability to do work. It is denoted by the letter '**E**.' Although in general energy is measured in Joules (J), electrical energy is typically expressed in Watt-hours (Wh) or kilo Watt hours (kWh). One kWh is equal to 1,000 Wh.

Power is the rate of doing work. It is denoted by the letter '**P**' and is measured in Watts (W).

Here is how the above are related:

The Ohm's Law:

$$I = V \div R$$

The Power Law:

$$P = I * V = I^2 * R = V^2 \div R$$

The Energy Law:

$$E \ [Wh] = P * t$$

Where **t** is time in 'hours.'

Appendix B: Getting the PSH from the NASA solar database

The **Perfect Sun Hours (PSH)** value is needed to calculate the solar-generated electricity. PSH can be determined by:

- Graphical solar maps, or

- Solar radiation database.

Graphical solar maps are the easiest to work with, but they are relatively inaccurate. Solar radiation datasets are both precise and reliable. The most popular online solar database is the NASA dataset.

In the NASA dataset, you can find the average annual PSH and the worst-month PSH of your site. Average annual PSH helps you calculate annual solar electricity produced by any solar system. The worst-month PSH helps you calculate the battery bank capacity for an off-grid solar panel system.

This is NASA solar database starting page:

https://eosweb.larc.nasa.gov/cgi-bin/sse/grid.cgi?

To access NASA database, you need to have an account:

You create your account by entering your name, email, and phone number:

After you are already registered in NASA database, you can access it by signing in with your email and password:

After you sign in you are lead to the 'Location' page where you are required to enter the coordinates of your location:

To get the needed PSH from a solar database, you need to know the coordinates of your location –

latitude and longitude. There are many ways to obtain the coordinates of your site, but one of the easiest one is searching Wikipedia:

http://en.wikipedia.org/wiki/Main_Page

If your site – whether a city, town, village or whatever settlement – has a page in Wikipedia, be sure you can find there its coordinates, both in degrees-minutes-second and in decimal degrees. If your location does not have a site in Wikipedia, you should use the coordinates of the nearest city or town instead.

Below you can see some screenshots showing how to get the coordinates of your location from Wikipedia.

First, you should enter the name of your location and click 'Enter':

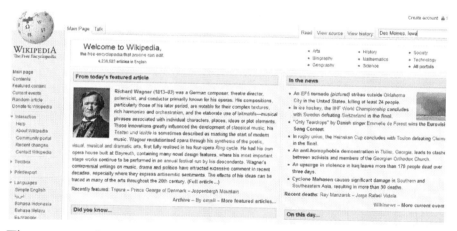

Then you have to search for its coordinates. They must be displayed somewhere on the page:

If you don't need latitude and longitude in degrees/minutes/seconds, but rather in decimal degrees, click on the coordinates and when you get to the next page, copy the coordinates in decimal degrees

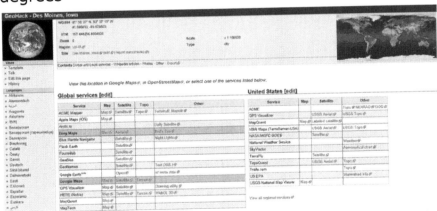

and enter them in the input fields on the 'Location' page:

After clicking the 'Submit' button, you go to the 'Choices' page where the only selection you have to make here is related to **Parameters for Tilted Solar Panels**:

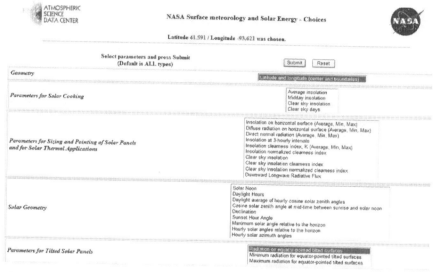

When you click on the 'Submit' button, you will be lead to the page containing the info on the location selected:

Latitude 41.591 / Longitude -93.621 was chosen.

Geometry Information

Elevation: 307 meters
taken from the
NASA GEOS-4
model elevation

Northern boundary
42

Western boundary
-94

Center
Latitude 41.5
Longitude -93.5

Eastern boundary
-93

Southern boundary
41

Parameters for Tilted Solar Panels:

Monthly Averaged Radiation Incident On An Equator-Pointed Tilted Surface (kWh/m²/day)

Lat 41.591 Lon -93.621	Jan	Feb	Mar	Apr	May	Jun	Jul	Aug	Sep	Oct	Nov	Dec	Annual Average
SSE HRZ	1.92	2.62	3.66	4.58	5.43	6.10	6.08	5.35	4.48	3.17	2.04	1.65	3.92
K	0.47	0.47	0.48	0.47	0.49	0.52	0.53	0.53	0.54	0.51	0.46	0.46	0.49
Diffuse	0.74	1.05	1.49	1.96	2.30	2.49	2.28	1.99	1.54	1.14	0.82	0.66	1.53
Direct	3.52	3.76	4.25	4.42	4.93	5.73	5.93	5.48	5.38	4.51	3.40	3.16	4.54
Tilt 0	1.91	2.56	3.62	4.56	5.39	6.13	6.02	5.32	4.41	3.14	2.02	1.64	3.90
Tilt 26	2.96	3.42	4.30	4.84	5.36	5.83	5.90	5.52	5.09	4.14	2.97	2.65	4.42
Tilt 41	3.35	3.68	4.40	4.68	4.99	5.33	5.43	5.26	5.13	4.42	3.31	3.03	4.42
Tilt 56	3.55	3.74	4.27	4.31	4.40	4.59	4.71	4.74	4.90	4.46	3.46	3.25	4.20
Tilt 90	3.26	3.15	3.23	2.84	2.62	2.59	2.67	2.94	3.50	3.67	3.10	3.04	3.05
OPT	3.58	3.75	4.40	4.84	5.47	6.13	6.08	5.55	5.15	4.47	3.47	3.29	4.69
OPT ANG	63.0	53.0	40.0	24.0	12.0	4.00	9.00	19.0	36.0	51.0	60.0	66.0	36.3

NOTE: *Diffuse radiation, direct normal radiation and tilted surface radiation are not calculated when the clearness index (K) is below 0.3 or above 0.8.*

Parameter Definition

You need the PSH averaged for the worst month. The worst-month averaged PSH is the lowest PSH among the PSH values from January till December for a given tilt. If 41° (Tilt 41) is the angle nearest to your roof slope, then the worst-month PSH is 3.03, and it occurs in December. Therefore, December is the month with the least solar radiation, so you can use PSH=3.03 to calculate the battery bank capacity. The optimal angle provided here will not do any work, since we are chasing the worst case.

If you want to be completely sure that the system will withstand your consumption needs during the worst month, you could choose the second option from the 'Choices' table *'Minimum radiation for equator-pointed tilted surfaces'*:

Select parameters and press **Submit**
(Default is ALL types)

Submit   Reset

Geometry	Latitude and longitude (Center and boundaries)
Parameters for Solar Cooking	Average insolation Midday insolation Clear sky insolation Clear sky days
Parameters for Sizing and Pointing of Solar Panels and for Solar Thermal Applications	Insolation on horizontal surface (Average, Min, Max) Diffuse radiation on horizontal surface (Average, Min, Max) Direct normal radiation (Average, Min, Max) Insolation at 3-hourly intervals Insolation clearness index, K (Average, Min, Max) Insolation normalized clearness index Clear sky insolation Clear sky insolation clearness index Clear sky insolation normalized clearness index Downward Longwave Radiative Flux
Solar Geometry	Solar Noon Daylight Hours Daylight average of hourly cosine solar zenith angles Cosine solar zenith angle at mid-time between sunrise and solar noon Declination Sunset Hour Angle Maximum solar angle relative to the horizon Hourly solar angles relative to the horizon Hourly solar azimuth angles
Parameters for Tilted Solar Panels	Radiation on equator-pointed tilted surfaces Minimum radiation for equator-pointed tilted surfaces Maximum radiation for equator-pointed tilted surfaces
Parameters for Sizing Battery or other Energy-storage Systems	Minimum available insolation as % of average values over consecutive-day period Horizontal surface deficits below expected values over consecutive-day period Equivalent number of NO-SUN days over consecutive-day period

Certainly, completely different results for PSH are displayed for a range of tilts equal to:

- 0°
- Latitude – 15°
- Latitude
- Latitude + 15°
- 90°
- The optimal tilt (the tilt at which the panel gets the maximum solar radiation possible)

Latitude 41.591 / Longitude -93.621 was chosen.

| Geometry Information | | Elevation: 307 meters
taken from the
NASA GEOS-4
model elevation |

Northern boundary
42

Western boundary
-94

Center
Latitude 41.5
Longitude -93.5

Eastern boundary
-93

Southern boundary
41

Parameters for Tilted Solar Panels:

Minimum Radiation Incident On An Equator-pointed Tilted Surface (kWh/m²/day)

Lat 41.591 Lon -93.621	Jan	Feb	Mar	Apr	May	Jun	Jul	Aug	Sep	Oct	Nov	Dec	Annual Average
SSE MIN	1.61	1.99	2.77	3.75	3.80	5.18	4.59	4.42	3.32	2.40	1.32	1.37	3.04
K	0.40	0.36	0.36	0.39	0.34	0.44	0.49	0.43	0.40	0.38	0.30	0.38	0.38
Diffuse	0.76	1.07	1.51	1.99	2.29	2.51	2.43	2.12	1.68	1.22	0.80	0.68	1.59
Direct	2.76	2.49	2.89	3.28	2.70	4.81	3.92	4.20	3.70	3.23	1.95	2.45	3.19
Tilt 0	1.60	1.94	2.74	3.73	3.77	5.39	4.56	4.40	3.27	2.38	1.31	1.36	3.03
Tilt 26	2.33	2.42	3.10	3.90	3.72	4.96	4.45	4.51	3.61	2.92	1.68	2.05	3.31
Tilt 41	2.59	2.55	3.12	3.76	3.49	4.55	4.13	4.29	3.59	3.05	1.80	2.30	3.27
Tilt 56	2.71	2.55	2.99	3.46	3.11	3.96	3.64	3.88	3.40	3.05	1.82	2.43	3.09
Tilt 90	2.45	2.10	2.24	2.32	1.98	2.35	2.20	2.47	2.44	2.44	1.56	2.23	2.23
OPT	2.72	2.56	3.13	3.91	3.81	5.20	4.59	4.54	3.62	3.06	1.82	2.45	3.46
OPT ANG	61.0	49.0	35.0	21.0	10.0	4.00	9.00	18.0	31.0	46.0	53.0	63.0	33.2

NOTE: *Diffuse radiation, direct normal radiation and tilted surface radiation are not calculated when the clearness index (K) is below 0.3 or above 0.8.*

Parameter Definition

If solar panels are to be installed at a fixed tilt (for example, on a sloped roof where it is rarely possible to change the tilt of the array), you have to select in the first column the tilt nearest to the tilt of your solar array. In case of sloped roof mounting, you should know the slope of your roof to compare it to the listed values. Having selected the nearest tilt to your roof slope, you should write down the lowest value among the twelve monthly averaged values, for the months from January till December.

For example, if your roof slope is 45°, your best option is to select the row of Tilt 41. For Tilt 41, the lowest monthly averaged PSH value is in November. It is 1.82, so November is the month with the least solar radiation for the selected location.

If you have the option of choosing your tilt, you should set the tilt of the solar array equal to the optimal tilt. In this example, the optimal tilt averaged on an annual basis for the location is 33.2° (the last cell in the last row in the table). Averaged annual PSH for the optimal tilt is 3.46 (in the above cell) which is much more favorable than 1.82. Higher PSH means a battery bank of lower capacity which is less expensive, smaller and easier to maintain.

Appendix C: Impact of irradiance and temperature

Here is a set of I-V curves each corresponding to a different value of irradiance.

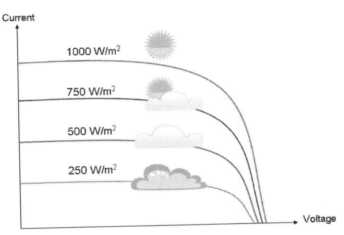

The current of a solar module is more affected by the irradiance than by the voltage. Higher irradiance means higher current and hence, higher power output. Lower irradiance means lower current and hence, lower power output. The other essential factor affecting the solar module performance is temperature. The performance of the solar module decreases as temperature increases.

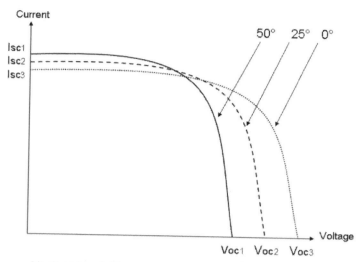

Voc3 and Isc3: Voc and Isc at T1=0° (32 F)
Voc2 and Isc2: Voc and Isc at T2=25° (77 F)
Voc1 and Isc1: Voc and Isc at T3=50° (122 F)

Appendix D: Temperature influence on solar panel performance

The CTI-160 solar panel specifications:

Rated power (Pm)	160W
Maximum power voltage (Vmp)	18.60V
Maximum power current (Imp)	8.60A
Open-circuit voltage (Voc)	22.50V
Short-circuit current (Isc)	9.29A
Power coefficient	−0.39% / °C
Voltage coefficient	−0.31% / °C
Current coefficient	0.045% / °C
Max power tolerance	+/- 5W

We are going to estimate the value of solar panel's parameters for lowest ambient temperature in Iowa - 15°C (6F).

Power:

$$P(T) = P(@25°C) + P(@25°C) * TKP * (T − 25 + + Ktemp),$$

Where:

- P(T) is the power at the ambient temperature T, W.

- P(@25°C) is the power at cell temperature 25°C, taken from the solar panel datasheet, W.
- TKP is the temperature coefficient of the power in %/°C, taken from the solar panel datasheet.
- T is the ambient temperature in degrees Celsius.
- Ktemp is a correction coefficient in degrees Celsius, accounting for the fact that the cell temperature is on average at least 15°C higher than the maximum positive ambient temperature. For negative temperatures, this coefficient is practically 0.

Example:

P(-15) = 160+160*(-0.39/100)*(-15–25+0) =
= 184.96W

Voltage:

V(T) = V(@25°C) +V(@25°C) * TKV *
* (T–25+Ktemp)

Where:

- V(T) is the voltage at the ambient temperature T, V.
- V(@25°C) is the voltage at cell temperature 25°C, taken from the solar panel datasheet, V.
- TKV is the temperature coefficient of the voltage in %/°C taken from the solar panel datasheet.
- T is the ambient temperature in degrees Celsius.
- Ktemp is a correction coefficient in degrees Celsius accounting for the fact that cell temperature is on

average at least 15°C higher than the maximum positive ambient temperature. For negative temperatures, this coefficient is practically 0.

Example:

We are going to calculate the open-circuit voltage Voc for -15°C:

Voc(-15) = 22.5+22.5*(-0.31/100)*(-15-25+0) =
= 25.29V

and for 38°C, assuming that Ktemp=15°C:

Voc(38) = 22.5+22.5*(-0.31/100)*
*(38-25+15)=20.55V

Current:

I(T) = I(@25°C) + I(@25°C) *TKI *
* (T−25+Ktemp)

Where:

- I(T) is the current at ambient temperature T.
- I(@25°C) is the current at cell temperature 25°C, taken from the solar panel datasheet.
- TKI is the temperature coefficient of the current in %/°C taken from the solar panel datasheet.
- T is the ambient temperature in degrees Celsius.
- Ktemp is a correction coefficient in degrees Celsius, accounting for the fact that the cell temperature is on average at least 15°C higher than the maximum positive ambient temperature. For negative temperatures, this coefficient is almost 0.

Example:

The short-circuit current Isc at -15°C is:

$$I(-15) = 9.29+9.29*(0.045/100)*(-15-25+0) =$$
$$= 9.12A$$

Source:

Pop, Lacho, Dimi Avram (2015-07-19). The Ultimate Solar Power Design Guide: Less Theory More Practice, Paperback Edition. Digital Publishing Ltd.

Appendix E: Guidelines for sizing an MPPT charge controller

You must ensure that the solar array will stay within the input maximum power point DC operational window of the charge controller, that is:

- Vocmax(Tmin)≤MPPT charge controller maximum input DC voltage

- Vocmin(Tmax) ≥ MPPT charge controller minimum input DC voltage

- Vmppmax(Tmin)≤ MPPT charge controller maximum MPP voltage

- Vmmpmin(Tmax) ≥ MPPT charge controller minimum MPP voltage,

Where Vocmin and Vocmax are the solar array maximum and minimum open-circuit voltages, and Vmppmax and Vmppmin are the maximum and minimum voltages at the maximum power point.

Tmax and Tmin are the maximum and minimum ambient temperatures.

Source:

Pop, Lacho, Dimi Avram (2015-07-19). The Ultimate Solar Power Design Guide: Less Theory More Practice, Paperback Edition. Digital Publishing Ltd.

References

1. Antony, Falk, Christian Durschner, Karl-Heinz Remmers. 2007. Photovoltaics for Professionals: Solar Electric Systems Marketing, Design and Installation, Routledge.

2. Hankins, Mark. 2010. Stand-Alone Solar Electric Systems: The Earthscan Expert Handbook for Planning, Design and Installation, Earthscan.

3. Mayfield, Ryan. 2010. Photovoltaic Design and Installation for Dummies, Wiley Publishing Inc.

4. Pop MSE, Lacho, Dimi Avram MSE (2015-02-17), The New Simple and Practical Solar Component Guide (Kindle Locations 1198-1199). Digital Publishing Ltd.

5. Pop MSE, Lacho, Dimi Avram MSE. (2015-10-26).The Truth About Solar Panels: The Book That Solar Manufacturers, Vendors, Installers And DIY Scammers Don't Want You To Read, Kindle Edition. Digital Publishing Ltd.

6. Pop MSE, Lacho,Dimi Avram MSE. 2015. The Ultimate Solar Power Design Guide: Less Theory More Practice, Kindle Edition. Digital Publishing Ltd.

7. Carmanah. CTI 160 Solar Module [Internet]. [cited 2017 Dec 31]. Available from: gpelectric.com/files/gpelectric/.../MOBI_SPC_CTI-160_vA.pdf

8. Go Power! INDUSTRIAL PURE SINE WAVE GP-ISW2000 / GP-ISW3000 [Internet]. 2016. Available from:

http://gpelectric.com/files/gpelectric/documents/PDF/SPEC_GP-ISW2000-3000.pdf

9. Morningstar Corporation. Morningstar's TrakStarTM MPPT Technology & Maximum Input Power - Morningstar Corporation [Internet]. [cited 2017 Dec 31]. Available from: https://www.morningstarcorp.com/whitepapers/morningstars-trakstar-mppt-technology-maximum-input-power/

10. Morningstar Corporation. Morningstar-ProStar-Datasheet [Internet]. Available from: https://www.morningstarcorp.com/products/prostar/

11. Morningstar Corporation. Morningstar Solar Controller System Sizing with 60 Cell Modules - Google Search [Internet]. [cited 2017 Dec 31]. Available from: http://support.morningstarcorp.com/wp-content/uploads/2014/07/tech-tip-60-cell-PV-module-sizing.pdf

12. Samlexamerica. 3000W 12VDC-120VAC Pure Sine Wave Power Inverter | SA-3000K-112 [Internet]. [cited 2017 Dec 31]. Available from: http://www.samlexamerica.com/products/ProductDetail.aspx?pid=129

13. VertaMax 3000 Watt 12V Pure Sine Wave Power Inverter DC to AC Car, RV, Solar, Off-Grid Applications [Internet]. [cited 2017 Dec 31]. Available from: https://www.windynation.com/Inverters/Windy-Nation/VertaMax-3000-Watt-12V-Pure-Sine-Wave-Power-Inverter-DC-to-AC-Car-RV-Solar-Off-Grid-Applications/-/1441?p=YzE9NDg=

Web resources:

http://photovoltaic-software.com/solar-radiation-database.php

https://eosweb.larc.nasa.gov/cgi-bin/sse/grid.cgi?

https://asdc-arcgis.larc.nasa.gov/sse/

https://play.google.com/store/apps/details?id=com.clarkgarrett.solartilt&hl=en

http://solarpanelsvenue.com/free-solar-cable-size-calculator/

http://openclipart.org

http://wikipedia.org

Index